MICRO AND NEURO-BIOLOGY

MICRO AND NEURO-BIOLOGY

Editor

G.C. Satpathy

KALPAZ PUBLICATIONS

DELHI-110052

MICRO AND NEURO-BIOLOGY

(Biology)

ISBN: 81-7835-054-6

Published in 2001 in India by
Kalpaz Publications
C-30, Satyawati Nagar
Ashok Vihar Phase-III
Delhi-110052
E-Mail: kalpaz@hotmail.com

Laser Typesetting at : A.S. Printers
Printed at : Mehra Offset Printers, New Delhi

Preface

Biology is a living science, in which changing knowledge continually generates fresh perspectives and fresh opportunities for productive impacts on our society. Our goal in this book is to provide you with the experimental basis of current understanding and to give you the tools to participate in the development of future knowledge. *Micro and Neurobiology* concentrates on studies on the fields of microbiology and neurobiology. All the concepts of this book is derived form scientific experiments.

The first section of this book deals with microbiology. Branch of biology which deals with the study of small forms of life is known as microbiology. Ecologists study microbiology to understand the natural recycling of minerals such as carbon and nitrogen. Evolutionists look to the microorganisms to learn more about the ancestors of contemporary forms, and biochemists use microorganisms as miniature laboratories to discover the chemical systems that underlie all living things. Geneticists observe hereditary material at work in microscopic forms and then apply what they have learned to more complex organisms. In this book the fundamentals of microbiology are presented within the framework of scientific discovery. Researchers in microbiology have made almost incredible strides in the past few decades. Consequently, existing concepts microbiology have expanded. There has been a revolution in this direction. The text integrates the descriptive, experimental and biochemical approaches into a conceptual framework for the analysis of microbiology.

The second section gives detailed reviews of emerging issues in neurobiology for which there are no comprehensive and current reviews available from other sources. We also continue our effort to provide reviews that span the continuum form molecular and cellular to behavioural approaches to neurobiology. All the chapters of neurobiology show the progress that has become increasingly important for analysis of brain function. We are now entering a more integrative phase of neurobiology in which molecular electrophysiological, and behavioural techniques must be combined to answer pressing questions about complex neuronal function.

Students of biology often have quite diverse academic backgrounds. Certainly introductory courses in biology and chemistry are prerequisite to a study of microorganisms and neurons. The present book is an attempt to make the students acquainted thoroughly with knowledge in both fields. It is prepared for students in biological, medical and health-oriented studies.

Editor

Contents

PART I

Microbiology

Chapter 1

Introduction to Microbiology

The branch of biology which deals with the study of small forms of life is known as microbiology. Ecologists study microbiology to understand the natural recycling of minerals such as carbon and nitrogen. Evolutionists look to the microorganisms to learn more about the ancestors of contemporary forms, and biochemists use microorganisms as miniature laboratories to discover the chemical systems that underlie all living things. Geneticists observe hereditary material at work in microscopic forms and then apply what they have learned to more complex organisms.

Microorganisms

Microorganisms exhibit numerous features consistent with those in more complex cells. For example, DNA is the hereditary material in all organisms, and many of the biochemical patterns of growth are identical. However, a number of significant differences do set microorganisms apart.

Before we explore these differences and unique features, we shall examine the total spectrum of living things by distinguishing two broad groups of organisms, the prokaryotes and eukaryotes. One of the most important generalizations to emerge in recent decades is that all organisms may be categorized as either prokaryotes or eukaryotes. Both terms are derived from the Greek word *karyon* meaning "nut," a reference to the nucleus of the cell. Prokaryotes lack a nucleus ("pro" implies primitive), while eukaryotes possess a well-defined nucleus ('eu" means true). In some texts, the words are spelled procaryote and eucaryote, but we shall retain the original spelling.

The distinctions between prokaryotes and eukaryotes go deeper than the nucleus. The cells of eukaryotes have a variety of membrane bound compartments called organelles in which important cellular functions take place. At some stage these organelles include mitochondria, Lysosomes, Golgi bodies, and an endoplasmic reticulum. Prokaryotic cells have none of these. In addition, eukaryotic cells have multiple chromosomes that together with protein make up a membrane-bound nucleus. Prokaryotic cells by contrast, have only a single chromosome composed solely of a molecule of DNA in a closed loop.

When eukaryotic cells contain chlorophyll pigments for photosynthesis, they occur in a distinct organelle, the chloroplast in prokaryotic cells that have chlorophyll, the pigments are dissolved in the cytoplasm. All prokaryotes and eukaryotes possess ribosomes, the RNA-protein bodies that participate in protein synthesis. Prokaryotic ribosomes are smaller than their counterparts in eukaryotic cells. They exist free in the cytoplasm, whereas eukaryotic ribosomes are bound to membranes. Furthermore, prokaryotic cells generally are surrounded by complex cell walls, whereas the cell walls of eukaryotic cells are either absent or have less complex chemistry when present.

Among the microorganisms, the protozoa, the fungi, and the unicellular algae are eukaryotes. By contrast, microorganisms such as the bacteria, rickettsiae, and cyanobacteria are prokaryotes. Knowing this has substantial practical value because we can define many of the properties of a microorganism merely by knowing whether its cells are eukaryotic or prokaryotic.

The following sections summarize other significant properties of the major groups of microorganisms.

Bacteria

Bacteria are prokaryotes and are among the most abundant organisms on Earth. The vast majority play a positive role in nature: they digest sewage into simple chemicals; they extract nitrogen from the air and make it available to plants for conversion to amino acids and proteins; they break down the remains of all that die and recycle the carbon; and they produce foods for our consumption and products for industrial technology.

Many biologists believe that life as we know it would be impossible without the bacteria. Of course, some bacteria are harmful. Certain bacteria multiply within the human body where they digest the tissues or produce poisonous toxins that lead to disease. Other bacteria infect plant crops or animal herds. Disease-causing bacteria are a global threat to all forms of life.

Bacteria have adapted to more different living conditions than any other group of organisms. They inhabit the air, soil, and water, and exist in enormous numbers on the surfaces of virtually all plants and animals. They can be isolated from Arctic ice, hot springs, the upper atmosphere, and the tissues of animals. Some can withstand the acid in vinegar, the crushing pressures of ocean trenches, or the enzyme activity of digestive juices.

Other bacteria survive in oxygen-free environments, boiling water, or extremely dry locations. Bacteria have so completely invaded every part of the earth that the mass of bacterial cells is estimated to outweigh the mass of all plants and animals combined.

Small bacteria

Rickettsiae and chlamydiae are prokaryotes. These microorganisms were considered a group unto themselves, but contemporary scientists classify them as bacteria and refer to them as "small bacteria." *Rickettsiae* are among the tiniest microorganisms. They are generally transmitted among humans by arthropods such as ticks and lice. They grow and multiply only within living cells and are responsible for such diseases as Rocky Mountain spotted fever and typhus fever. *Chlamydiae* are a subgroup of rickettsiae. They are among the smallest rickettsiae and they cause human diseases of the eyes, lungs, and urogenital tracts. Another type of bacteria, mycoplasmas, dwarf the chlamydiae. Although prokaryotes, they do not possess cell walls.

Cyanobacteria

Cyanobacteria were known as blue-green algae. Today they are considered more closely related to bacteria than to algae because of their structural and biochemical properties. However, cyanobacteria are not regarded as true bacteria, because of fundamental differences in the process by which they form carbohydrates in photosynthesis. Cyanobacteria are prokaryotes. They possess light-trapping pigments that function in photosynthesis taking place in the cell. Many of the pigments are blue, but some are black, yellow, green, or red. The periodic redness of the Red Sea, for example, is due to a species of cyanobacteria that contains large amounts of red pigment.

Cyanobacteria may occur as unicellular or filamentous organisms. Many species can convert atmospheric nitrogen to organic compounds useful to plants. Cyanobacteria inhabit freshwater as well as marine environments. When ponds or lakes contain a rich supply of nutrients, cyanobacteria may "bloom" and convert the water to a pea-soup green with a foul taste. Swimming pools and aquaria may experience this problem if algicide is not used regularly.

Protozoa

Protozoa are single-celled eukaryotes. Many thousands of species exist, among which about two dozen species cause diseases such as malaria and sleeping sickness.-The word "protozoa" is derived from Latin stems that mean " first-animals," a term reflecting one view of their evolutionary significance. Protozoa exhibit a bewildering assortment of shapes, sizes, and structural components. A few species have the necessary pigments for photosynthesis and can therefore synthesize their own foods. Most species, however, must obtain nutrients from preformed organic matter.

Protozoa generally exist wherever there is water and a supply of food. The majority of species are microscopic, although a few may be seen with the unaided eye. Protozoa are classified into groups according to their means of locomotion. Some have whiplike flagella, others possess hairlike cilia, and still others move by means of cytoplasmic extensions called pseudopodia (false-feet).

Fungi

Fungi are eukaryotic organisms that include two groups, the yeasts and molds. Yeasts are unicellular organisms about the size of large bacteria. They play a vital role in the fermentation of wine and beer and the production of bread. Molds are long chains of cells often seen as fuzzy masses. Commonly they assume vivid colours due to the pigments in spores they use for reproductive purposes. Fungi lack photosynthetic pigments, and so they must obtain preformed nutrients from the environment. Together with bacteria, fungi are the primary decomposers of organic matter in the world.

Algae

Algae are plantlike organisms that practice photosynthesis and differ structurally from ordinary land plants such as mosses, ferns, and seed plants. Two types of unicellular algae, the diatoms and the dinoflagellates, are significant in microbiology.

Diatoms

Diatoms are eukaryotic marine microorganisms that serve as a main source of food in the world's oceans. Through photosynthesis, they trap the sun's energy and form carbohydrates that are passed on to other marine organisms as food. The cell walls of diatoms are impregnated with silicon, a glasslike substance. When diatoms die, their remains accumulate on the seashore as diatomaceous earth that is gathered and used to produce bacterial filters. Dinoflagellates are photosynthetic eukaryotes important in food chains as well as cause's of red tides. Viruses are neither prokaryotes nor eukaryotes.

Many microbiologists even question whether viruses are living organisms because they are non-cellular. They do not grow nor do they show any nutritional patterns. Viruses have no observable activity except replication, and they accomplish this function only within living cells. Viruses consist essentially of a fragment of nucleic acid, either DNA or RNA, surrounded by a sheath of protein.

Although we shall consider them "microorganisms," viruses appear to be at the threshhold of life. Despite their simplicity, viruses are responsible for many human diseases including influenza, hepatitis, polio, and chickenpox.

Classification

All microorganisms have names, and all have certain size limitations. In addition, there are classification schemes into which most fit. In this section we shall explore the principles on which these properties are based. The science of classification, or taxonomy, deals with the systematized arrangements of related microorganisms and other living things into logical categories. Derived from Latin roots meaning "law of arrangements," taxonomy is essential to an understanding of the relationships among living things. It focuses on unifying concepts among organisms while providing a basis for communication among biologists.

In the fourth century Aristotle categorized living forms in the world around him by describing over 500 species according to their appearance and habits. Without reference books or instruments to guide him, Aristotle made some key observations, including the fact that the whale was more like a mammal than a fish.

The modern basis of taxonomy was devised by the Swedish botanist Carl von Linne, better known to history as Carolus Linnaeus. Linnaeus named thousands of plants and animals and classified them in the kingdoms Plantae and Animalia. He inspired an unprecedented worldwide program of specimen hunting and sent his students around the globe in search of new life forms. He also devised the binomial scheme of nomenclature, thus resurrecting the use of Latin as a language of the learned.

In his otherwise exacting system, Linnaeus reflected the scant knowledge of microorganisms and his general disinterest in them by grouping them separately under the heading of Vermes (as in vermin) in the category Chaos (confusion). In the years after Linnaeus, however, some microorganisms came to be considered plants while others were categorized as animals.

Ernst H. Haeckel separated the microorganisms from the plant and animal kingdoms. Haeckel coined the term protista for microorganisms and placed the fungi, protozoa, and microscopic algae in a third kingdom, Protista. As the bacteria were discovered they were included in this group. However, biologists could not fully agree on the placement of bacteria and fungi.

Advances in the study of cell biology and increasing interest in evolutionary biology raised serious questions about two-or three-kingdom classification schemes, during the twentieth century. Robert H. Whittaker of Cornell University proposed a system that has gained wide acceptance among modern biologists.

Further expanded in succeeding years by Lynn Margulis of Boston University, the system recognizes five kingdoms for all living things: Monera, Protista, Fungi, Animalia, and Plantae. According to the five-kingdom system, the bacteria and cyanobacteria belong to the kingdom Monera because they are the only true prokaryotes.

The second kingdom, Protista, includes the protozoa and unicellular algae. These eukaryotic organisms are unicellular all their lives; many contain chlorophyll (plantlike), and move (animal-like).

The fungi are placed in the third kingdom, Fungi, because they are nongreen eukaryotic organisms whose cell walls differ chemically from those of traditional plants. Also the cytoplasm mingles among adjacent cells in the fungus, so that fungi are not true multicellular organisms. The final two kingdoms, Plantae and Animalia, are the traditional plants and animals. Note that the viruses are not to be found in any of these kingdoms because their characteristics fit neither the prokaryotes nor the eukaryotes. Viruses appear to be partial or degenerate forms of living systems.

Mechanics of the classification system have remained consistent since first outlined by Linnaeus. The fundamental rank is the species (pl., species), a group of microorganisms that have overall similarity and differ significantly from other species. Two or more species are grouped together as a genus (pl., genera). A collection of genera make up a family, and families with similar characteristics make up an order. Orders are placed together as a class, and classes are assembled into a phylum.

A microorganism may often belong to a rank below the species level to indicate that a special characteristic exists within a subgroup of the species. Such ranks have no official standing in nomenclature, but often they have practical usefulness in helping further to pinpoint an organism.

For example, two biotypes of the cholera bacillus, *Vibrio cholerae,* are known to exist: *Vibrio cholerae* classical biotype and *Vibrio cholerae El* tor biotype. Other designations of ranks include subspecies, serotype, strain, morphotype, and variety. Some bacteriologists recommend changing all these subspecies ranks to varieties and then using names such as biovar and serovar.

Nomenclature

All microorganisms have a double name, usually from Latin or-Greek stems. The name consists of the genus to which the organism belongs and a species modifier that further describes the genus name. This system of nomenclature is called the binomial system from the Latin bi-nomiall meaning two names. First suggested by Linnaeus, the binomial system gives biologists throughout the world an international language for life forms and eliminates incalculable amounts of confusion. When the genus and species modifier of an organism are written, only the first letter of the genus is capitalized. The remainder of the genus name and species modifier are written in lowercase letters.

Both words should be printed in italics but if this is not possible, they should be underlined. For example, the species of intestinal rod discovered by Theodor Escherich is written as *Escherichia coli* or Escherichia coli. Scientists often abbreviate binomial species names by writing the first letter of the genus, or some accepted substitution, together with the full species modifier. The abbreviation should also be italicized or underlined. Thus *Escherichia coli* becomes *E-coli, and Bacillus subtilis* is written as *B. subtilis.*

Size

Microorganisms are so small that special units are employed. The unit of measurement most frequently used is the micrometer. This is equivalent to a millionth (micro-) of a meter. Scientists symbolize the micrometer as μm, that is, a combination of the Greek letter "μ" (pronounced "mue") and the abbreviation "m," for meter. A micrometer is also equivalent to a thousandth of a millimeter. Microorganisms range in size from the relatively large, almost visible protozoa (100 μm) down to the incredibly tiny viruses (0.0 1 μm), ten thousand times smaller.

Molds consist of intertwined filaments so long and twisted that they are visible, but the individual mold cells measure only about 40 μm by 10 μm. One group of fungi, the yeasts, are commonly about 8 μm in diameter. Though most bacteria measure about 1 μm to 5 μm in length, the largest ones reach approximately 20 μm in length while the smaller ones, the rickettsiae, may be only about 0.5 μm long. Chlamydiae, among the smallest bacteria, are a scant 0.25 μm in length.

Because the viruses are so tiny, scientists often express their size in nanometers to avoid decimals of micrometers. The nanometer symbolized as nm, is equivalent to a billionth (nano-) of a meter. The size of a smallpox virus may therefore be written as 250 nm, rather than 0.25 μm. The polio virus, among the smaller viruses, measures 20 nm. Other objects measured in nanometers include the wavelength of radiant energy such as visible or ultraviolet light, and the size of certain large molecules.

Modern science have made available to microbiologists a broad range of instruments for viewing the microorganisms. These instruments include the common light microscope, a, number of specialized instruments, and the highly sophisticated electron microscopes. All operate on the same basic principle: energy is projected toward an object, such as a microorganism. The energy bounces off the object and creates an impression on a sensing device. This device may be a television screen, a photographic film, or the human eye. The image reveals the form, shape, size, and other structural features of the object.

Microscopic technologies

Basic microscopic system used in the microbiology laboratory is the light microscope. It is also called the bright-field microscope because visible light passes directly through its lenses until it reaches the eye. Another common term for this instrument is the compound microscope because it has a two-lens system, with the objective lens nearer the object and the ocular lens nearer the eye.

Light microscopy

In light microscopy, visible light is projected through a substage condenser, which focuses the light into a sharp cone. The light then passes through the opening in the stage, into the slide, and bounces off the object. Next, it enters the objective lens to form a magnified image darker than the background. This image is called a real image because it can be projected onto a screen.

However, the image is not seen by the microscopist. Instead, the image becomes an object for the ocular lens, which magnifies the image a second time to create a virtual image in space. Only the observer can see this image. It is about as distant from the eye as this page is from your eye.

The light microscope usually has three objective lenses, the low-power, high-power, and oil-immersion lenses. Generally these magnify an object about 10, 40, and 100 times, respectively. The magnification is represented by the multiplication symbol " x." Total magnifications are determined by multiplying the magnification achieved by the objective by the magnification achieved by the ocular. With a 10 X ocular, the total magnifications achieved would be 100 X, 400 X, and 1000 X, respectively. In order for an object to be seen distinctly, the lens system must have good resolving power, that is, it must transmit light without variation and allow nearby objects to be clearly distinguished.

In preparation for light microscopy, microbiologists commonly stain bacteria because the cytoplasm of bacteria lacks colour. Several techniques have been developed for this purpose. To perform the simple stain technique, a small amount of bacteria is placed in a droplet of water on a glass slide and the slide is air-dried. Next the slide is passed through a flame in a process called heat-fixing. This bonds the cells to the slide, kills many organisms that may still be alive, and prepares them for staining. Now the slide is flooded with a basic dye such as crystal violet or methylene blue. Cytoplasm generally has a negative charge and since basic dyes have a positive charge, the dye is attracted to the cytoplasm where staining takes place. The negative stain technique works in the opposite manner. Bacteria are mixed on a slide with an acidic dye such as nigrosin, a black stain, or congo red, a red dye. The mixture is then smeared across the face of the slide and allowed to air-dry. Because the acidic dye carries a negative charge, it is repelled by the cytoplasm. The stain gathers around the cells and the microscopist observes clear or white cells on a coloured background.

Since this technique avoids chemical reactions and heat-fixing, the cells appear less shriveled and less distorted, and are closer to their natural condition. The Gram stain technique allows us to view stained cells while learning something about them. It is named for Christian Gram, the Danish physician who first suggested its use.

The procedure is a differential technique because it differentiates bacteria into two groups depending upon the results. Certain bacteria arc called Gram-positive bacteria; others are Gram-negative. The first two steps of the technique are straightforward. Air-dried, heat-fixed smears are stained with crystal violet, then with a special Gram's iodine solution. All bacteria become blue-purple. Next the smear is rinsed with a decolorizer such as 95 percent alcohol or an alcohol-acetone mixture. At this point, certain bacteria lose their colour and become transparent. These are the Gram-negative organisms. Other bacteria retain the blue-purple stain. These are the Gram-positive bacteria. When safranin, a red dye, is applied to the slide next, only the Gram-negative organisms will accept the stain. Thus at the technique's conclusion, Gram-positive bacteria are blue-purple while Gram negative organisms appear red. By observing the colour of the cells at the conclusion of the process, one may decide the group to which the bacteria belong.

One theory suggests that crystal violet and iodine form a chemical complex in the bacterial cytoplasm. Since Gram-negative bacteria have a high lipid content in their cell walls, some microbiologists maintain that the alcohol dissolves the lipid and allows the crystal violet-iodine complex to leak out-of the cytoplasm. Gram-positive bacteria, with less cell-wall lipid, are less susceptible to the alcohol's effects.

Another theory points to the heavy concentration of peptidoglycan in the cell wall of Gram-positive bacteria. Peptidoglycan, a complex carbohydrate, is thought to trap the crystal violet-iodine complex in its many cross linkages. Gram-negative bacteria have considerably less peptidoglycan in their cell walls and hence would trap less of the complex. It should be noted that the terms "positive" and "negative" are nothing more than convenient expressions and that electrical charges play a minimal role in Gram staining. Knowing whether an organism is Gram-positive or Gram-negative is important for several reasons. For instance, microbiologists use results from the Gram stain technique to identify an unknown organism and classify it in *Bergey's Manual.* Gram-positive and Gram-negative bacteria differ in their susceptibility to chemical substances such as antibiotics, and they have different structural components. They produce different types of toxic poisons as well.

Dark-field microscopy

The dark-field microscopy is having a background which always remains dark, and only the object is illuminated. A special condenser mounted under the stage scatters the light and causes it to hit the object from different angles. Some light bounces off the object into the lens to make the object visible, but the surrounding area appears dark because it lacks background light. Dark-field microscopy helps in the diagnosis of diseases caused by spiral bacteria because these organisms are near the limit of resolution.

For example, syphilis is due to *Treponema pallidum*, a spiral bacterium with a diameter of about 0. 15 μm. This bacterium may be observed in scrapings taken from the lesion of one who has the disease. The unstained organisms are seen moving about with a characteristic rotary motion.

Phase-contrast microscopy

In phase-contrast microscopy, special condensers are also used. A series of special condensers and filters split a light beam and throw the light rays slightly out of phase. The separated beams of light then pass through and around microscopic objects, and small differences in the densities of the objects show up as different degrees of brightness and contrast. The Dutch scientist Fritz Zernicke received the Nobel Prize in Physics for his development of the system. With the phase-contrast microscope microbiologists can see organisms alive and unstained. Internal details not visible with the light microscope may also be observed.

Fluorescent microscopy

Fluorescent microscopy has emerged to become a major asset to the diagnostic and research laboratory. The technique has been applied to the identification of many microorganisms and is a mainstay of modern microbiology. In fluorescent microscopy, microorganisms are coated with a fluorescent dye, such as fluorescein, and illuminated with ultraviolet light energy. The energy excites electrons in the dye and they move to higher energy levels. However, they quickly drop back to their original energy levels and give off the excess energy as visible light. The coated microorganisms appear to fluoresce.

An important application of fluorescent microscopy is in the flulorescent antibody technique used to identify an unknown organism. In one variation of this procedure, fluorescein is chemically united with antibodies, the protein molecules produced by the body's immune system when stimulated by a specific organism. "Tagged" antibodies result. Next, these antibodies are mixed with a sample of the unknown organism. If the antibodies are specific for that organism they will bind to it and coat the cells with the dye. When subjected to ultraviolet light, the organisms will fluoresce.

Electron microscopy

Electron microscope grew out of an engineering discovery made by German scientists. They found that a flow of electrons would occur between opposite poles in a scaled tube if they maintained a vacuum to prevent wandering of the electrons. Magnets pinpointed the flow onto an object where the electrons were absorbed or deflected, depending on the density of structures within the object. When projected onto a screen underneath, the electrons formed an image that outlined the structures.

American engineers at RCA research laboratories had adapted the device to an instrument for microscopy and were publishing photographs with magnifications of 10,000 times. The key to electron microscopy is the extraordinarily short wave-length of the beam of electrons, Measured at 0.005 nm, the short wavelength increases the resolving power of the system to fractions of a nanometer and makes possible the visualization of viruses, fine cellular structures, and large molecules such as DNA. Two types of electron microscopes are currently in wide use.

The first type, the transmission electron microscope (TEM) is used to photograph detailed structures within cells. Ultrathin sections of the object must be prepared because the electron beam can penetrate matter only a short distance. After embedding the specimen in a suitable mounting medium or freezing it, scientists cut the specimen into sections with a diamond knife. In this manner, a single bacterium can be sliced the long way into a hundred or more sections.Once sectioned, the portions of the object are floated in water, and picked up on a wire grid. Then they are stained with a heavy metal such as gold or palladium to make certain parts dense. Next, the microscopist inserts the sections into the vacuum chamber of the instrument and focuses a 10,000-volt electron beam on them using magnetic lenses. An image forms below. A photograph prepared from the image may be enlarged with enough resolution to achieve a total magnification of over 20 million times.

The second type is the scanning electron microscope. This instrument, developed in the late permits the surfaces of objects to be seen in the natural state, without sectioning. The specimens are placed into the vacuum chamber and covered with a thin coat of gold to increase electrical conductivity and decrease blurring. The electron beam then sweeps across the object and knocks loose showers of electrons that are captured by a detector. An image builds line by line much as in a television receiver. Electrons that strike a sloping surface yield fewer electrons, thereby producing a darker contrasting spot and a sense of three dimensions.

Magnifications with the scanning electron microscope are limited to about 75,000 to 100,000 times, but the instrument is relatively easy to operate and gives vivid and undistorted views of an organism's surface details.

Development of microbiological studies

In England, a scientist named Robert Hooke published a major work called *Micrographic.* The book contained a miscellany of his thoughts on chemistry as well as a description of the microscope and its uses. Hooke did not invent the microscope, but he did make it widely known. Hooke's writings awakened the learned of Europe to the world of very small objects and creatures. Among his illustrations were representations of the eye of a fly, stinger of a bee, structure of a feather, and plantlike form of a mold.

Hook described a slice of cork and suggested that the cork was composed of compartments, or cells. This account secured for him a place in the history of cell biology. Other scientists were quick to follow Hooke's lead. For example, the Dutch naturalist Jan Swarnmerdam described tiny bodies, the red blood cells, in samples of blood. Another Dutchman, Regnier de Graaf, discovered structures in the animal ovary that are known today as Graafian follicles. Still another scientist, the Italian physiologist Marcello Malpighi, described the tiny capillaries of the animal cardiovascular system.

Discoveries by Hooke, Swammerdam, and other scientists showed that the microscope was an important tool for unlocking the secrets of nature. It is not surprising, therefore, that genuine interest was forthcoming when Anton van Leeuwenhoek revealed his descriptions of microorganisms. In his spare time, he ground pieces of glass into fine lenses, placing them between two silver or brass plates riveted together. Later he added an adjustable device for holding tiny specimens. Van Lecuwenhoek's lenses were no larger than the head of a pin, but they served him well. By most accounts they could magnify an object over 200 times. Initially he used the lenses to inspect the quality of cloth, but as his fascination with microscopic objects developed, he examined hair fibers, blood cells, and even samples of his own feces.

At one point, van Leeuwenhock observed tiny sperm cells and speculated that they contained microscopic ernbryos that were transferred to the female for nourishment and development. Van Leeuwenhock's work did not go unnoticed or unreported. Probably the only scientific group of that period was the Royal Society of London, but correspondence with this group was tenuous because England and the Netherlands were bitter rivals for commercial treasures of the East Indies. They did so, and soon van Leeuwenhoek was sending along his illustrations and observations.

The contact was maintained for the next 50 years. Van Leeuwenhoek filled a glass with greenish, cloudy water from a marshy lake outside of Delft, and placed a sample under his lens. The water teemed with tiny microorganisms, which he called animalcules. His curiosity aroused, van Leeuwenhoek soon located the animalcules in rainwater, material from his own teeth and feces, and eventually, most of the specimens he examined. The creatures astonished him at first, then delighted him, and finally, perplexed him as he pondered their origin and purpose.

Van Leeuwenhock sent his eighth letter to the Royal Society. Van Leeuwenhoek's sketches were elegant in detail and clarity. In succeeding letters, he outlined structural details of the familiar protozoa known today as *Paramecium* and *Amoeba,* and described the threadlike fungi, as well as certain microscopic algae. A particularly noteworthy letter, included drawings of what are thought to be bacteria in the form of rods, spheres, and spirals.

Van Leeuwenhock was elected to fellowship in the Royal Society and, with Isaac Newton and Robert Boyle, he became one of the most famous men of his times. Peter the Great of Russia and the Queen of England came to peer into his microscopes. His technique was so fastidious that when some of his slides were discovered 300 years later, scientists found perfectly preserved specimens of cotton seeds, slices of cork, and the optic nerve of a cow.

In all, van Leeuwenhoek bombarded the Royal Society with over 200 letters about his findings. He died at the age of 90, his advanced age itself an achievement. He was a very suspicious and secretive person. He invited no one to work with him, nor did he show anyone how to grind lenses or construct a microscope. This is one reason why interest in the microorganisms waned after his death. Another reason, and perhaps of more significance, is that scientists of the day saw the microorganisms merely as idle curiosities of nature. Disease was still shrouded in magic and mysticism, and few people believed that microorganisms could cause disease in so lofty a creature as the human being. Thus, the substantial development of microbiology would be delayed until the technology for an efficient microscope emerged and until an association was drawn between microorganisms and disease.

The science of microbiology blossomed during a period of about 60 years referred to as the golden age of microbiology.

The work of Louis Pasteur and continued into the twentieth century until the advent of World War I. During these years, numerous branches of microbiology were established and the foundation was laid for the maturing process that has led to modern microbiology. In a world ravaged by plague, tuberculosis, typhoid fever, and diphtheria, a large family was often necessary to ensure the next generation. Neither royalty nor commonfolk were immune to disease. There were virtually no cures for disease. Indeed, no one was really sure what caused it. Such were the times in which *Louis* Pasteur studied at the

French school, *Ecole Normale Superieure.* He achieved distinction in organic chemistry for his discovery that tartaric acid, a four-carbon organic compound, forms two different types of crystals.

Pasteur successfully separated the crystals while looking through a microscope. In doing so, he developed a skill with that instrument that aided his later studies of microorganisms. At the age of 32, he was appointed Professor of Chemistry at the University of Lille in northern France. Pasteur had a deep patriotism for France. He also believed that the discoveries of science should have practical applications.

Almost immediately, Pasteur stirred controversy. Popular theory held that fermentation was a natural process in which protein in grape juice broke down to alcohol in wine. However, Pasteur's microscope consistently revealed large numbers of oval yeast cells, and occasionally a variety of barely visible sticks and rods known to scientists as bacteria. Other chemists had overlooked the yeast cells, but Pasteur believed that they played a major role in fermentation.

Pasteur showed that wine could be produced by incubating grape juice with yeasts, but not by incubating protein with yeasts. This indicated that protein could not be the source of alcohol in wine. Next he destroyed all the yeasts in a sample of grape juice by heating the juice. When the sample was then set aside to ferment, Pasteur showed that nothing would take place. However, when he returned the yeasts to the grape juice, the fermentation proceeded as normal. His further experiments showed that if special care was taken to eliminate the bacteria, the wine would not turn sour. Pasteur's work shook the scientific community. His results demonstrated that yeast cells and bacteria were tiny, living factories where important chemical changes were taking place. His work also drew attention to microorganisms as the agents of change, because bacteria appeared to make the wine "sick." For years, bacteria had been interpreted as an effect of disease; that is, they were thought to arise in the body during illness.

Pasteur's work appeared to indicate that they could be a cause of disease, for if they could sour the wine, perhaps they could also make the body ill. Pasteur wrote a short paper relating microorganisms to illness and set the foundation for the germ theory of disease. Essentially this theory holds that microorganisms are responsible for infections diseases. Pasteur also recommended a practical solution to the sour wine problem. He suggested that grape juice be heated to destroy all the evidence of life, after which yeasts could be added to begin the fermentation. An alternative was to heat the wine before the bacteria soured it. Acceptance of his technique, known as pasteurization, gradually ended the problem.

Pasteur's interest in microorganisms rose as he learned more about them. He found bacteria in soil, water, air, and the blood of disease victims. Extending his germ theory of disease, he reasoned that if microorganisms were acquired from the environment, then their spread could be controlled. If so, then perhaps the chain of disease transmission could also be broken. However, many scientists stubbornly stuck to the notion that bacteria arose spontaneously from organic matter and that disease was inevitable so long as the "life force" was present.

Pasteur was therefore drawn into the lingering debate on spontaneous generation, and he sought to discredit the doctrine in order to salvage his own theory of disease.

Pasteur first showed that where disease was rampant, the air was full of microorganisms, but where the air was clean, disease was uncommon. He opened flasks of broth to air from the crowded city, then from the countryside, and next, from a high mountain. In each succeeding experiment, fewer flasks became contaminated with microorganisms. Still, his critics were vocal. When he boiled broths and showed that they remained free of life, the critics argued that boiling destroyed the "life force."

Finally, in an elegant series of experiments, Pasteur silenced all but the most extreme supporters of spontaneous generation. He prepared broth in a series of swan-neck flasks, so named because their S-shaped necks resembled a swan's. He left the flasks open to the air and any "life force." However, the S-shaped curvature of the neck trapped dust and microorganisms and prevented their entry to the flask. The flasks of broth were boiled, then left undisturbed, No microorganisms grew in the broth. When the neck was later cut off, though, airborne organisms quickly fell into the broth and growth appeared within hours.

Pasteur's work brought to an end the long and tenacious debate on spontaneous generation that had begun two centuries earlier. By now he was a national celebrity. Once more, however, tragedy entered his life; his daughter Camille died of blood poisoning. Pasteur realised that he was no closer to solving the riddle of disease.

That same year, cholera engulfed Paris, killing 200 people a day. Pasteur attempted to capture the responsible bacteria by filtering the air in hospital wards through cotton. He obtained a variety of bacteria but was unable to cultivate one type of microorganism apart from others in the mixture. Although he demonstrated that bacterial inoculations made animals ill, he could not pinpoint an exact cause. Some of his critics claimed that a poison, or toxin, in the broth was responsible for the disease.

In an effort to help French industry, Pasteur turned his attention to pebrine, the disease of silkworms. Late in 1865, he identified a protozoan infesting the sick silkworms and the mulberry leaves fed to them. Next he separated the healthy silkworms from the diseased silkworms and their food, and managed to quell the spread of disease. The achievement strengthened the germ theory of disease.

For Pasteur, however, it was another time of grief. Cecille, another of his daughters, died of typhoid fever. Again he returned to the study of human disease. Though Pasteur failed to relate a specific organism to a specific disease, his work stimulated others to investigate the nature of microorganisms and to ponder their association with disease. For example, Gerhard Hansen, the Norwegian physician, identified bacteria in the tissues of leprosy patients and Otto H. F. Obermeier of Germany described bacteria in the blood of victims of relapsing fever. Another German bacteriologist, Ferdinand J. Cohn discovered that bacteria multiply by dividing into two cells. He also observed that certain ones form an extremely resistant structure called the spore.

In England, Joseph Lister was sufficiently impressed with Pasteur's writings on disease to begin applying antiseptics to wound infections. However, the definite proof of the germ theory of disease was offered by Robert Koch a country doctor from East Prussia, now part of Germany. Koch's primary interest was anthrax, a deadly blood disease in cattle and sheep.

Anthrax was a threat to farmers because it ravaged their herds periodically and seemed to reappear time and again in the same district without warning. Casimir Davaine, a French physician, had cultivated a bacillus from diseased animals. Koch was determined to learn more about anthrax. A makeshift laboratory in his home he injected mice with the blood of diseased sheep and cattle. He then performed meticulous autopsies and noted how the same symptoms appeared regularly. Next, he isolated a few rod-shaped bacilli from a mouse's blood by placing the bacilli in the sterile vitreous humor from an ox's eye.

Koch watched for hours as the bacilli multiplied, formed tangled threads, and finally reverted to the resistant spores described by Cohn. At this point, he took several spores on a sliver of wood and injected them to healthy mice. The symptoms of anthrax appeared within hours. Koch autopsied the animals and found their blood swarming with bacilli. He reisolated the bacilli in sterile vitreous humor. The cycle was now complete. Koch communicated his findings to Cohn, Cohn invited the 35-year-old physician to present his work at the University of Breslau. Scientists there were astonished. Here was the proof for the germ theory of disease that had eluded Pasteur and for which many of them were waiting.

Koch's procedures came to be known as Koch's postulates. Quickly were adopted as a guide for relating specific organisms to specific diseases. They stated that in order to relate a microorganism to a disease, the microorganism must be identified in the affected organism, the microorganism must be isolated and grown in pure culture, and the disease must be reproduced in experimental animals by these pure cultures. The same microorganism must then be recovered.

Originally the postulates had been outlined by Jakob Henle, a German researcher. After the sensation at Breslau subsided, Koch returned to his laboratory and developed staining methods for bacteria. He accepted an appointment to the Imperial Health Office and during this period he happened upon the cultivation techniques that sparked the further development of microbiology. Koch chanced to observe that a slice of potato contained small masses of bacteria, which he termed colonies. Colonies contained millions of just one kind of bacteria.

Koch concluded that bacteria could grow and multiply on solid surfaces, and he added gelatin to his broth to prepare a solid culture medium. He then inoculated bacteria to the surface and set the medium aside to incubate. Next morning visible colonies were present on the surface. When colonies of the same bacillus grew together, a pure culture formed. Koch could now inoculate laboratory animals with a pure culture of bacillus and be certain that only one type of bacterium was involved. His work also proved that bacteria, not toxins in the broth, were the cause of disease. Koch demonstrated his pure culture techniques to the International Medical Congress, meeting at Lister's laboratory in London. Pasteur and several of his co-workers were present as

Pasteur and Koch might have become friends and colleagues in their mutual search for the agents of disease. Franco-Prussian war were characterized by fierce national pride. Both France and Germany were undergoing unification, and heroes played an important role in the spirit of nationalism. Thus, Pasteur became a symbol of French achievement and Koch, his German counterpart. There sprung up a rivalry that was to last into the next century. Koch's proof of the germ theory was presented in 1876.

Within two years, Pasteur had verified the proof and gone a step further. He reported that bacteria were temperature-sensitive because chickens did not acquire anthrax at their normal body temperature of 42°C but did so when the animals were cooled down to 37°C. He also recovered anthrax spores from the soil and pointed out that cattle were probably infected during grazing. This explained the periodic recurrence of the disease. Pasteur suggested that dead animals be burned or buried deeply in soil unfit for grazing. One of Pasteur is more remarkable discoveries occurred when a group of inoculated chickens failed to develop chicken cholera. Even after a second injection with fresh and deadly bacilli, the animals remained healthy.

Pasteur then discovered that the first injection was with "old" bacilli grown for a long time above normal incubation temperatures. In a brilliant moment of insight, he reasoned that the first inoculation made the animals resistant to disease. Thus, modified microorganisms could be used to protect animals against deadly forms. This principle is the basis for the use of many vaccines in immunity. Pasteur applied the principle to anthrax and found he could protect sheep against the disease. Experiments such as these put France once more in the forefront of science.

However, Koch's announcement of pure culture techniques drew attention back to Germany, and within a year, Koch isolated the tubercle bacillus, the cause of tuberculosis. His, associate, Georg Gaffky, cultivated the typhoid bacillus, and that same year, another co-worker, Friederich Loeffler, isolated the diphtheria bacillus. Soon, news was forthcoming from France that Emile Roux and Alexandre Yersin of Pasteur's group had proven diphtheria to be caused by a toxin that the bacillus produced in the body.

In later years, Koch's co-worker, Emil von Behring, was the first to successfully treat diphtheria by injecting antitoxin. This blood product is obtained from animals injected with diphtheria toxin. Von Behring was awarded the first Nobel Prize in Physiology or Medicine. There was also an international flavor in the French and German laboratories. Shibasaburo Kitasato of Japan studied with Koch and, in and muscle damage in successfully cultivated the tetanus bacillus, an organism that adults grows only in the absence of oxygen.

One of Pasteur's associates was Elie Metchnikoff, a native of the Ukraine. Metchnikoff was the first to publish an account of *phagocytosis*, a defensive process in which certain white blood cells of the body engulf and destroy microorganisms in the blood. This period also witnessed two improvements in microscopy, both attributed to the German physicist Ernst Karl Abbe. Abbe introduced the oil immersion lens that is a standard feature of modern microscopes.

Eight years later he invented the system of lenses and mirrors known as the Abbe condenser. This apparatus concentrates light upon objects being viewed and makes detailed vision of microorganisms feasible. Thus, the technology for seeing microorganisms dovetailed the interest in microorganisms. Pasteur reached the zenith of his career when he successfully immunized young Joseph Meister against the dreaded disease rabies.

Although he never saw the rabies organism, Pasteur was able to cultivate it in the brain of animals and inject the boy with samples of the tissue. The experiment was a triumph for Pasteur because it fulfilled his dream of applying the principles of science to practical problems. Many monetary rewards were also forthcoming, including a generous gift from the Russian government after Pasteur immunized 20 Russians against rabies. The funds were used to help establish the Pasteur Institute in Paris, one of the world's foremost scientific establishments.

Koch's isolation of the tubercle bacillus was part of a continuing interest in tuberculosis for the remainder of his years. He interrupted his work to lead groups studying cholera in Egypt and India. In both countries, Koch isolated a comma-shaped bacillus and confirmed the suspicion first raised by John Snow 30 years earlier that water transmitted the disease. He was appointed to the University of Berlin and he became Director of the Institute for Infectious Diseases.

With the aging of Pasteur and Koch, a new generation of scientists stepped in to expand their work. For example, a Pasteur Institute scientist, Charles Nicolle, proved that typhus fever was transmitted by lice. Albert Calmette, also of the Institute, developed a harmless strain of the tubercle bacillus that could be used for immunization. Jules Bordet, another French scientist, isolated the bacillus of pertussis, and developed the complement fixation test, a procedure widely used in the diagnosis of disease. Koch's successors included Emil von Behring, and Richard Pfeiffer, who isolated one of several organisms that cause meningitis. Still another co-worker was Paul Ehrlich, a chemist who explored the mechanisms of immunity and eventually synthesized the "magic bullet," a sulfur compound that would seek out and destroy syphilis organisms in the human body.

Microbiology had moved far beyond the boundaries of France and Germany. For instance, Ronald Ross, an English physician working in the Far East, proved that the mosquitoes were the vital link in malaria transmission. The discovery earned him the Nobel Prize. Another Englishman, David Bruce, isolated the cause of undulant fever. While working in Africa, Bruce also showed the relationship between tsetse flies and sleeping sickness. The control of this insect opened the African continent to British colonization. A third British subject, Almroth Wright, described opsonins, the chemical substances that assist phagocytosis in the body. Thirty years later, one of his students, Alexander Fleming, discovered penicillin. Two Japanese investigators also achieved distinction in these years. The Tokyo physician, Masaki Ogata, reported that rat fleas were the agent of transmission in bubonic plague epidemics. This discovery solved a centuries old mystery of how plague spread. A year later, Kiyoshi Shiga isolated the bacterium that causes dysentery, an important intestinal disease. The organism was later named *Shigella.*

Advent of World War I signaled a dramatic pause in microbiology research and brought to an end the golden age of microbiology. The Pasteur Institute was closed as Paris came under siege, and the German laboratories focused on production of antibacterial scrums for treating war-related diseases. The golden age witnessed a series of discoveries unparalleled in medicine, most involving identification of the agents of disease. Although cures for established diseases would not come until the scientists developed an awareness that disease was caused by microorganisms and that the chains of transmission could be broken. Their discoveries led to calls for sterile practices in hospitals, pasteurization of milk, purification of water, control of insects, and care in the preparation of food for consumption.

Together with improved sanitation and personal hygiene, these measures brought about a substantial reduction in the incidence of bacterial disease. However the agents of such diseases as measles, mumps, smallpox, yellow fever, and polio continued to elude microbiologists. The belief developed that some sort of "virus" was to blame and that discovery of the virus was simply a matter of time. Nevertheless, microbiologists found a way to immunize against their effects and interrupt their transmission. Many viral diseases are coming under control much as many bacterial diseases did at the turn of the century. Smallpox has not occurred in the world and polio, measles, and mumps are on the rapid decline. As this trend continues to take place, we shall be watching for the diseases of the future.

Chapter 2

Microbial Diseases

In this chapter we shall discuss the complex mechanisms that underlie the spread and development of microbial diseases. We shall summarize much of the important terminology used in medical microbiology and focus on the methods used by microorganisms to establish themselves in the tissues. The word infection is derived from Latin origins meaning to mix with or corrupt. The term refers to the relationship between two organisms, the host and the parasite, and the competition for supremacy that takes place between them. A host that gains the upper hand remains healthy, and the parasite is either driven from the host or assumes a benign relationship with the host. By contrast, if the host loses the competition, disease develops.

The term disease appears to have originated from Latin stems that mean living apart, a reference to the separation of ill individuals from the general population. Disease may be conceptualized is any change from the general state of good health. It is important to note that disease and infection are not synonymous; a person may be infected without becoming diseased. The concept of infection in the host-parasite relationship is expressed in the body's normal flora. The normal flora is a population of microorganisms that infect the body without causing disease. Some organisms in the population establish a permanent relationship with the body, while others are present for limited periods of time.

Relationship between the body and its normal flora is an example of a symbiosis. In some cases the symbiosis is beneficial to both the body and the microorganisms. This relationship is called mutualism. For example, species of *Lactobacillus* live in the human vagina and derive nutrients from the environment while producing acid to prevent the overgrowth of other organisms. In other cases, the symbiosis is beneficial only to the microorganisms, in which case the symbiosis is called commensalism.

Escherichia coli is generally presumed to be a commensal in the intestine of the body, although some evidence exists for mutualism because certain amounts of vitamins B and K are derived from the bacteria. A normal flora may be found in several of the body tissues. On the skin, for instance, one may locate various forms of viruses, fungi, and bacteria, particularly staphylococci and *Propionibacterium acnes.* The oral cavity commonly contains members of the genera *Neisseria, Leptotrichia,* and *Bacteroides*, as well as many diphtherialike bacilli (diphtheroids), fungal spores, and streptococci. The upper respiratory tract is also the site of these organisms, as well as pneumococci and species of *Haemophilus* and *Mycoplasma.* Organisms such as these may cause serious respiratory diseases if the body defenses are compromised during cases of influenza.

The stomach contents in humans are without a normal flora mainly due to the low pH of its contents. However, the latter part of the small intestine and the large intestine abound with microorganisms. *Bacteroides* species are numerous, together with *Clostridium* spores, various streptococci, and a number of Gram-negative rods including species of *Enterobacter, Klebsiella, Proteus,* and *Pseudomonas. Escherichia coli* is it well-known resident of the intestine, as is *Candida albicans*, the yeast. *Lactobacillus* is a notable component of the normal flora in the vagina; other organisms may be located near the urogenital orifices in both males and females. The blood and urine are usually sterile.

Organisms of the normal flora are introduced when the child passes through the birth canal. Additional organisms enter when breathing begins and upon first feeding. Within two to three days most organisms of the flora have appeared. During the next few weeks, contact with the mother and other individuals will expose the child to additional microorganisms. The normal flora remains throughout life, undergoing changes in response to the internal environment of the individual.

Pathogenicity

The word "pathogenicity" is derived from the Greek term pathos, meaning suffering. The term pathogen has the same root and refers to an organism having pathogenicity. The symbiotic relationship between host and parasite is called parasitism. Parasites vary greatly in their pathogenicity. Certain parasites such as the cholera, plague, and typhoid bacilli are well known for their ability to cause serious human disease. Other organisms, such as common cold viruses, are considered less pathogenic because they induce milder illnesses. Still other organisms are opportunistic. These organisms may be commensals in the body until the normal defenses are suppressed, at which time the commensals seize the "opportunity" to invade the tissues and act as pathogens.

Streptococcus pneumoniae, for example, commonly resides on the surface of the upper respiratory tract without incident, but should the tissue be damaged, it may invade the tissue and cause pneumonia. Another example is observed in individuals with acquired immune deficiency syndrome (AIDS). These patients are highly susceptible to opportunistic organisms such as *Pneumocystis carinii, Toxoplasma gondii,* and species of *Cryptosporidium*. None of these organisms was considered a serious pathogen before the emergence of AIDS. Diseases caused by opportunistic organisms illustrate how a shift in the body's delicate balance of controls may convert infection to disease. When antibiotics came into widespread use. As bacteria in the intestine's normal flora disappeared, the biological controls exerted on Candida albicans vanished and candidiasis became common. Antibiotic destruction of the lactobacilli in the vaginal tract also contributed to the establishment of candidiasis.

Today, a number of new organisms are entering the United States from foreign lands and are assuming commensal relationships with Americans. A chance upset in resistance mechanisms may enhance the ability of these newcomers to establish disease. A sobering concept has emerged in recent years. Traditionally, microbiologists believed that organisms were either pathogenic or nonpathogenic. This distinction has been blurred by the realization that normally benign organisms may be pathogenic when body defenses weaken or fail. The AIDS situation points up this concept well.

Moreover, scientists have shown that some "harmless" organisms are pathogenic in transplant patients and cancer victims treated with radiation and immune-suppressing drugs. It is now recognized that pathogenicity is a function of the aggressive nature of the parasite as well as the level of resistance in the host. The word virulence is often used to express The degree of pathogenicity of a parasite. This term is derived from the Latin *virulentus*, meaning full of poison. An organism such as the typhoid bacillus that invariably causes disease is said to be "highly virulent." By comparison, an organism such as *Candida albicans* that sometimes causes disease is labeled "moderately virulent."

Certain organisms, described as virulent are not generally regarded as disease agents. The lactobacilli and streptococci found in yogurt are examples. However, it should be noted that any microorganism has the ability to change, genetically and become virulent. *E. coli,* for example, was long considered an avirulent commensal of humans, but certain toxin-producing strains are now isolated in outbreaks of human diarrhea and urinary tract diseases.

Progress of disease

The disease begins with a period of incubation reflecting the time that elapses between entry of the parasite to the host and the appearance of symptoms. Incubation periods may be a short one to three days as in cholera, a moderate two weeks as in chickenpox, or a long three to six years as in leprosy. Such factors as the number of parasites, their generation time and virulence, and the level of host resistance determine the period's length. Also the location of entry may be a determining factor. For instance, the incubation period for rabies may be as short as several days or as long as a year depending on how close to the central nervous system the viruses enter the body.

The second phase in disease is the period of prodromal symptoms. This period is characterized by general symptoms such as nausea, fever, headache, and malaise, which indicate that the competition for supremacy has begun. The period of acme follows. This is the acute stage of the acute stage of the disease, when specific symptoms appear. Examples are the skin rash in scarlet fever, jaundice in hepatitis, teardrop-shaped vesicles in chickenpox, and swollen glands in infectious mononucleosis. Often, physicians observe high fever and chills, the latter reflecting differences in temperature between the superficial and deep areas of the body. Dry skin and pale expression may result from constriction of the skin's blood vessels to conserve heat. As the symptoms subside, the period of decline, or defervescence.

This period may be preceded by a crisis period after which recovery is often rapid. In other cases the period may last a long time. Sweating is common as the body releases excessive amounts of heat, and the normal skin colour soon returns as the blood vessels dilate. The sequence comes to a conclusion during the period of convalescence. During this time the body's systems return to normal.

There are clinical or subclinical diseases . A clinical disease is one in which the symptoms are apparent, while a subclinical disease is accompanied by few obvious symptoms. Many people, for example, have experienced subclinical cases of mumps or infectious mononucleosis and have developed immunity to future attacks.

By contrast, certain diseases are invariably accompanied by clearly recognized clinical symptoms. Measles and malaria are examples. Disease usually results from the growth and multiplication of parasites in the host. However, certain diseases are due to the ingestion of toxins.

Disease transmission

Agents of disease may be transmitted by a broad variety of methods conveniently divided into two general categories: i) direct methods and ii) indirect methods.

Direct methods

Direct methods of transmission imply close or personal contact with one who has the disease. Hand-shaking, kissing, sexual intercourse, or contact with feces are examples. Such diseases as gonorrhea and genital herpes, and numerous diseases of the intestine are spread by direct contact. Direct contact may also mean exposure to droplets. the tiny particles of mucus expelled from the respiratory tract during a cough or sneeze. Diseases spread by this method include influenza, measles, pertussis (whooping cough), and streptococcal sore throat. For some diseases, direct contact with an animal is necessary. Rabies, leptospirosis, and toxoplasmosis are typical.

Indirect methods

Indirect methods of disease transmission include consumption of contaminated food or water, as well as contact with fomites. Foods are contaminated during processing or handling, or they may be dangerous when made from diseased animals. Poultry products, for example, are often a source of salmonellosis because *Salmonella* serotypes frequently infect chickens; and pork may spread trichinosis because the parasites live in muscles of the pig. Fomites are inanimate objects that carry disease organisms. For instance, bed linens may be contaminated with pinworm eggs, and contaminated syringes may carry the viruses of hepatitis B.

Another indirect method for transmission is seen in arthropods. Living organisms, such as arthropods, that carry disease agents from one host to another are called vectors. In some cases the arthropod may be a mechanical vector of disease because it transports microorganisms on its legs and other body parts. In other cases, the arthropod itself is diseased and serves as a biological vector. In malaria and yellow fever, for instance, disease organisms infect the arthropod and accumulate in its salivary gland, from which they are injected during the next bite.

In Rocky Mountain spotted fever and Lyme disease, microorganisms gather in the tick's gut and enter a wound when the tick feeds in human skin. For disease to perpetuate itself a continuing source of disease organisms in nature is necessary. These sources are called reservoirs. In smallpox, for example, the sole reservoir of viruses is humans. Thus by locating all humans who had the disease, the World Health Organization was able to limit the spread of the virus and eradicate smallpox from the earth. Meningococcal meningitis and diphtheria also have humans as reservoirs.

Disease carrier

A carrier is one who has recovered from the disease but continues to shed the disease agents. For instance, people who have recovered from typhoid fever or amebiasis are usually carriers for many weeks after the symptoms of disease have left. Their feces may spread the disease to others via contaminated food or water. Animals may also be reservoirs of disease. Domestic housecats, for example, usually show no symptoms of toxoplasmosis but are able to transmit *Toxoplasma gondii* to humans, where the disease manifests itself, Water and soil may likewise be considered reservoirs since they are often contaminated with disease agents.

Certain communicable diseases are further described as contagious, because they pass with particular case among hosts. Chickenpox, measles, and genital herpes fall into this category. One reason for easy transmission is that the focus of disease is on or close to the body surface. Noncommunicable diseases are singular events where the agent is acquired directly from the environment and is not easily transmitted to the next host. In tetanus, for example, penetration of *Clostridium* spores to the dead anaerobic tissue of a wound must occur before this disease develops.

Types of diseases

Diseases may also be categorized as endemic diseases, epidemic diseases, or pandemic diseases. An endemic disease is one that occurs at a low level in a certain geographic area. By comparison, epidemic diseases break out in explosive proportions within a population, and pandemic diseases occur worldwide. Often the episode of disease is simply referred to as an endemic, epidemic, or pandemic, respectively. The dictionary of medical microbiology contains numerous terms that describe types of disease. For example, such words as "communicable" or "noncommunicable" are commonly applied to a disease to denote its ability to be transmitted; and "endemic" and "epidemic" indicate how widespread the disease is. The words "acute" and "chronic" are applied to diseases as relative measurements of their severity. An acute-disease develops rapidly, is usually accompanied by severe symptoms, comes to a climax, and then fades rather quickly.

Cholera, epidemic typhus, and yellow fever are examples of acute diseases. Chronic diseases, by contrast, often linger for long periods of time. The symptoms are slower to develop, a climax is rarely reached, and convalescence may continue for several months. Hepatitis A, trichomoniasis, and infectious mononucleosis illustrate chronic diseases. Sometimes an acute disease may become chronic when the body is incapable of ridding itself completely of the parasite. For example, one who has had giardiasis or amebiasis may experience this problem. Infections may develop as primary or secondary diseases. A primary disease occurs in an otherwise healthy body while a secondary disease develops in a weakened individual. Influenza pandemic, hundreds of millions of individuals contracted influenza and then developed pneumonia as a secondary disease. Most of the deaths in the pandemic were due to complications such as this.

Local diseases are restricted to a single area of the body, while systemic diseases are those that disseminate to the deeper organs and systems. Thus a staphylococcal skin boil beginning as a localized skin lesion may become more serious when staphylococci spread and cause systemic disease of the bones, meninges, or heart tissue. Streptococci spreading from the gum tissues may also be hazardous.

The word commonly used for dissemination of bacteria through the bloodstream is bacteremia. Another term, septicemia, which means an infection spreading in the blood, is often used as a synonym. The "septicemia" or "blood poisoning" appearing in older textbooks is now regarded as streptococcal or staphylococcal blood disease. Fungemia refers to the spread of fungi, viremia to spread of viruses, and parasitemia to spread of protozoa and multicellular worms to the blood. We have used the word "disease" loosely to define a clinical condition initiated by an infectious microorganism. It should not be inferred, however, that this description is true of all diseases.

Establishment of disease

The parasite must possess unusual abilities if it is to overcome host defenses and bring about the anatomical or physiological changes leading to disease. Before it can manifest these abilities, however, the parasite must first gain entry to the host in sufficient numbers to establish a population. Next, it should be able to penetrate the tissues and grow at that location. Disease is therefore a complex series of interactions between parasite and host.

Portal of entry of the parasite

'Portal of entry' refers to the site at which the parasite enters the host. It varies considerably for different organisms and is a key factor in the establishment of disease. For example, tetanus may occur if *Clostridium tetani* spores are introduced to the anaerobic tissue of a wound but not if they are consumed in food because the spores do not germinate in the human intestine. This is why one can cat a freshly picked radish without fear of tetanus. Similarly, the bacilli that cause typhoid fever will not grow in the lung tissue, nor will the diplococci pneumonia flourish in the intestine.

If the sites of infection were reversed, however, disease might occur. One reason offered for specific portals of entry is the presence of adhesive factors on the surface of parasites. Certain parasites have multiple portals of entry. The tubercle bacillus, for instance, may enter the body by respiratory droplets, contaminated food and milk, and skin wounds. The Q fever organism enters by all these methods as well as by an arthropod bite. The tularemia bacillus may enter the eye by contact, the skin by an abrasion, the respiratory tract by droplets, the intestine by contaminated meat, or the blood by an arthropod bite. For certain arthropodborne diseases, entry must occur directly into the bloodstream.

Dose

Dose is the number of parasites that must be taken into the body in order for disease to be established. The consumption of one million typhoid bacilli in contaminated water will probably lead to disease. By contrast, many times that number of cholera bacilli must be ingested if cholera is to be established. One explanation offered is the high resistance of typhoid bacilli to the acid conditions in the stomach in contrast to the low resistance of cholera bacilli. Often the host is exposed to low doses of a parasite and as a result, develops immunity.

For instance, many people can tolerate low numbers of mumps viruses without exhibiting disease. They may be surprised to find that they are immune to mumps when it breaks out in their family at some later date. Other individuals have developed immunity to fungal pathogens after several low-grade exposures. Cases of histoplasmosis, for example, are regarded as a mild "summer flu" in the Ohio and Mississippi valleys, but histoplasmosis can be serious in an individual not previously exposed to the disease. It should be noted that immune people remain healthy because they enjoy strong resistance, again pointing to the notion that resistance and disease are fundamentally inseparable.

Tissue penetration

Knowledge of tissue penetration has been developed from studies on histological preparations, tissue cultures, and animals, and an understanding of the penetration process is generally incomplete. Experiments reported by Stanley Falkow and his co-workers at Stanford University indicate that genes for cell penetration exist on the chromosome of certain bacteria. These genes appear to code for surface proteins that assist penetration.

Falkow's group successfully isolated the penetration genes from *Yersinia pseudotuberculosis* and inserted them into *E. coli* which then displayed penetration. Other researchers have reported genes for penetration on the plasmids of *Shigella flexneri*. The ability of a parasite to penetrate tissues and cause structural damage is called *invasiveness*. This constitutes one of the major mechanisms by which a parasite may cause disease. The bacilli of typhoid fever and the amebas that cause amebiasis are well known for their invasiveness.

By penetrating the tissue of the gastrointestinal tract, these organisms cause ulcers and sharp, appendicitislike pain characteristic of the respective diseases. In some cases, penetration may not be critical to disease. The Pertussis bacillus, for example, remains on the Surface layers of the respiratory tract while producing the toxins that lead to disease.

Enzymes

Virulence of a parasite depends to some degree on its ability to produce a series of enzymes directed at host defenses. The enzymes act on specific cells and interfere with certain functions meant to retard invasion.

An example of a bacterial enzyme is the coagulase produced by virulent staphylococci. Coagulase catalyzes the formation of a blood clot from fibrinogen proteins in the human blood. The clot sticks to staphylococci, forming a mass of bacteria protected from phagocytosis. Part of the walling-off process observed in a staphylococcal skin boil is due to the clot formation. Coagulase-positive staphylococci are generally considered virulent. They may be identified in the laboratory by combining staphylococci with human or rabbit plasma. Formation of a clot in the plasma indicates coagulase activity. Streptococci have the ability to produce the enzyme streptokinase. This substance dissolves fibrin clots used by the body to restrict and isolate an infected area. Streptokinase thus overcomes an important host defense and permits further tissue invasion by the parasites.

Hyaluronidase is sometimes called the spreading factor because it enhances penetration of a parasite through the tissues. The enzyme digests hyaluronic acid, a polysaccharide that binds cells together in a tissue. The term "tissue cement" is occasionally applied to this polysaccharidc.

Hyaluronidase is an important virulence factor in pneumonococci and certain species of streptococci and staphylococci. In addition, gas gangrene bacilli use it to facilitate spread through the muscle tissues. This activity increases the severity of disease by contributing to invasiveness.

Leukocidin is a product of staphylococci, streptococci, and certain bacterial rods. The enzyme disintegrates circulating neutrophils and tissue macrophages, both of which are active phagocytes. Usually disintegration occurs before phagocytosis has taken place, but on some occasions it occurs after the phagocyte has engulfed the parasite. Hemolysins are a series of enzymes that dissolve red blood cells. Studies show that hemolysins combine with the membranes of erythrocytes, after which lysis takes place. In gas gangrene, hemolysins lead to substantial anemia. Staphylococci and streptococci are also known to Produce these virulence factors.

Toxins

The microbial poisons which profoundly affect the establishment and course of disease are known as toxins. Two types of toxins are recognized: exotoxins and endotoxins.

Exotoxins

Exotoxins are produced chiefly by grampositive bacteria. They are protein molecules, manufactured during the metabolism of bacteria. Exotoxins are released into the surrounding environment of the tissue as they are produced. They dissolve in the blood fluid and circulate to their site of activity. The symptoms of disease soon develop. The exotoxin produced by the botulism bacillus *Clostridium botulinum* is among the most lethal toxins known. One pint of the pure toxin is believed sufficient to destroy the world's population.

In humans, the toxin inhibits the release of acetylcholine at the synaptic junction, a process that leads to the paralysis in botulism. Another exotoxin is produced by tetanus bacilli. In this case, the exotoxin inhibits the removal of acetylcholine in the synapse, thereby stimulating volleys of spontaneous nerve impulses and uncontrolled muscle contractions. A third toxin is produced by Corynebacterium diphtheriae, the diphtheria bacillus.

The exotoxin interferes with the assembly of proteins in the cytoplasm of epithelial cells of the upper respiratory tract. Disintegrated cells then accumulate with mucus, bacteria, fibrous material, and white blood cells to cause life-threatening respiratory blockages.

Other exotoxins are formed by the bacteria that cause scarlet fever, staphylococcal food poisoning, pertussis, and cholera. When toxins function in a particular organ system, they are given more clearly defined terms. For example the botulism toxin is called a neurotoxin because of its activity in the nervous system, while the staphylococcal toxin is called an enterotoxin since it functions in the gastrointestinal tract. The body responds to exotoxins by producing special antibodies called antitoxins.

When toxin and antitoxin molecules combine with each other, the toxin is neutralized. This process represents an important defensive measure in the body. Therapy in people who have botulism, tetanus, or diphtheria often consists of injections of antitoxins to neutralize the toxins. Antibiotics for the bacteria are also used. Because exotoxins are proteins, they are susceptible to the heat and chemicals that normally react with proteins.

A chemical such as formaldehyde may be used to alter the toxin and destroy its toxicity without hindering its ability to elicit an immune response in the body. The result is a toxoid. When the toxoid is injected to the body, the immune system responds with antitoxins that circulate and provide a measure of defense to disease. Toxoids are used for diphtheria and tetanus immunizations in the diphtheria-pertussis-tetanus (DPT) vaccine.

Endotoxins

Endotoxins are part of the cell wall of bacteria and as such, are released only upon disintegration of the parasite. They are present in many Gram-negative bacilli and are composed of lipid-polysaccharideProtein complexes.

Endotoxins do not appear to stimulate an immune response in the body, not can they be altered to prepare antitoxins. Endotoxins manifest themselves by similar signs and symptoms. Usually an individual experiences an increase in body temperature, substantial body weakness and aches, and general malaise.

Damage to the circulatory system and shock may also occur. In this case, the permeability of the blood vessels changes and blood leaks into the intercellular spaces, where it is useless. The tissues swell, the blood pressure drops, and the patient may lapse into a coma. This condition, commonly called endotoxin shock, may accompany antibiotic treatment of diseases due to gramnegative bacilli since endotoxins are released as the bacilli disintegrate. Endotoxins usually play a contributing rather than a primary role in the disease process. Certain endotoxins have been observed to reduce

Platelet counts in the host and thereby hinder clot formation. Other enodotoxins are known to increase hemorrhaging.

Mocrobial diseaces and human host

Circulatory system

The blood circulatory system of the human host serves as the principal vehicle for the dissemination of parasites and their toxins. Many important factors in the host defense systems also operate within the circulation. These two concepts add to the importance of a basic understanding of the circulatory system and its capabilities.

There are three major components of the blood: the fluid, the clotting agents, and the cells. The fluid portion, called serum, is an aqueous solution of minerals, salts, proteins, and other organic substances. When clotting agents such as fibrinogen and prothrombin are present, the fluid is referred to as plasma. Three types of cells circulate in the blood: the red blood cells, 01 erythrocytes; the white blood cells, or leukocytes; and the platelets, or, thrombocytes. Erythrocytes arise in the bone marrow and carry oxygen to the tissues loosely bound to the red pigment hemoglobin.

A normal adult has about 5 million erythrocytes per cubic millimeter (mm^3) of blood. After circulating for about 120 days, erythrocytes disintegrate in the spleen, liver, and bone marrow. The hemoglobin is then converted to bilirubin, a pigment that gives bile a deep yellow colour. Normally bilirubin is carried to the liver for degradation in a protein-bound form. However, if the liver is damaged by a disease, or if too many erythrocytes disintegrate, excess bilirubin may enter the bloodstream, This condition, called jaundice, is responsible for the yellow colour of the complexion during cases of hepatitis and yellow fever.

White blood cells, or leukocytes, have no pigment in their cytoplasm and therefore they appear gray when unstained. These cells are also produced in the bone marrow. They number about 5000 to 9000 per mm' of blood and have different life spans depending on the type of cell.

One type of leukocyte has a multilobed nucleus and is therefore referred to as a polymorphonuclear cell, or PMN. In its cytoplasm it has many lysosomes that contain digestive enzymes. The cell is also called a neutrophil, because its granules stain with neutral dyes.

Neutrophils function chiefly as phagocytes. They pass out of the circulation through pores in the vessels and squeeze into narrow passageways among the cells to engulf particles. Approximately 55 to 60 percent of the leukocytes are neutrophils. Their life span is about 12 hours.

Basophils are types of leukocytes whose granules stain with basic dyes. The basophils number about 50 to 90 per mm^3 of blood and represent about 1 percent of the total number of leukocytes. They function in allergic reactions as their granules release physiologically active substances such as histamine.

Eosinophils are the third type of leukocyte, representing about 2 percent of the total number. These cells exhibit red cytoplasmic granules when an acidic dye such as eosin is applied. Substances in the granules arc thought to neutralize the active chemicals in basophil granules. Eosinophils are also postulated to have phagocytic activity.

Another major phagocyte of the circulatory system is the monocyte. This cell has a single, bean-shaped nucleus that takes up most of the area of the cytoplasm. Monocytes lack granules and account for approximately 5 to 8 percent of the leukocytes. In the tissues, monocytes exist as a type of phagocyte called the macrophage. In contrast to the two-week life span of monocytes, macrophages may live for several months.

Another type of leukocyte is the lymphocyte. This cell is produced in the bone marrow and transported to the lymph nodes. It has a single, large nucleus and no granules.

Lymphocytes make up about 30 to 35 percent of the white blood cells in the human body. They function in the immune system. Their numbers increase dramatically during the course of certain diseases, such as infectious mononucleosis.

Blood platelets, or thrombocytes, represent the third type of cell in the circulatory system. They are small disc-shaped cells that originate from cells in the bone marrow. Platelets have no nucleus and function chiefly in the blood clotting mechanism.

Lymphatic system

Lymph is the fluid that surrounds the tissue cells and fills the intercellular spaces is called tissue fluid, or. Lymph is similar to serum except that it has fewer proteins. Lymph bathes the body cells, supplying oxygen and nutrients while collecting wastes. It is pumped along in tiny vessels by the contractions of skeletal muscle cells. Eventually the tiny lymph vessels unite to form large vessels that compose a lymphatic system. The system empties into a large vein just before the heart. Pockets of lymphatic tissue located along the lymph vessels are known as lymph nodes. Lymph nodes are prevalent in the neck, armpits, and groin. They are bean-shaped organs containing phagocytes that engulf particles in the lymph and lymphocytes that respond specifically to substances in the circulation. Since resistance mechanisms are closely associated with the lymph nodes, it is not surprising that they become enlarged during time of disease. The tonsils, adenoids, spleen, Peyer's patches of the intestine, and appendix are specialized types of lymph nodes.

Bacterial diseases of humans

Airborne bacterial diseases

Streptococcal diseases

Streptococci are a large and diverse group of encapsulated gram-positive bacteria. They cause streptococcal sore throat, scarlet fever, and several other diseases in humans. The bacteria divide in one plane and thus cling together to form chains of various lengths. This observation is apparent in streptococci grown in liquid media, but it may not be obvious in streptococci from solid media. Microbiologists classify the numerous types of streptococci by several systems of which two are widely accepted. The first system, developed by J.H. Brown divides streptococci into hemolytic groups, depending on how they affect red blood cells. For example, *alpha-hemolytic streptococci* turn blood agar an olive-green colour as they partially destroy red blood cells in the medium. By contrast, colonies of *beta-hemolytic streptococci* are surrounded by a clear, colourless zone due to the complete destruction of red blood cells; and *gamma-hemolytic streptococci* have no effect on red blood cells and thus cause no change in blood agar.

The second classification system, suggested by Rebecca Lancefield is based on variants of a specific carbohydrate, the C substance, in cell walls of streptococci. Groups A through O streptococci have been distinguished. Most streptococcal diseases in humans are caused by species of Group A streptococci, one of which, *Streptococcus pyogenes*, is the most common. This beta-hemolytic organism is the one generally meant when physicians refer to "Group A beta-hemolytic streptococci." *Streptococcus pyogenes* is the cause of streptococcal sore throat, popularly known as *strep throat*. *S. pyogenes* is one of the few bacteria that can invade the normal tissues of the pharynx and cause a *primary disease*, that is, one developing in an otherwise healthy body.

The streptococci reach the upper respiratory tract within airborne *droplets*, the tiny particles of mucus expelled during coughing and sneezing. After a short incubation period (one to three days), a strep throat develops. Patients experience a high fever, swollen lymph nodes and tonsils, and a fiery red "beefy" appearance to pharyngeal tissues. About 500,000 Americans suffer strep throat annually. The pathogenicity of *Streptococcus Pyogenes* is enhanced by a substance called the *M protein*. This protein, located in the cell wall and pili, encourages adherence to the pharyngeal tissue and retards phagocytosis. Over 60 specific types of M protein have been identified, and complete immunity to streptococcal disease requires that a person produce antibodies against all 60 types. Thus an individual may suffer multiple bouts of strep throat over the course of a lifetime.

Treatments of streptococcal diseases is generally successful with antibiotics such as penicillin and erythromycin. However, widespread resistance to antibiotics has been noted and research indicates that many strains of *S. pyogenes* carry bacteriophages in a lysogenic relationship. The transfer of these phages among streptococci was demonstrated in study, and the movement of antibiotic-resistance genes was shown to parallel the transfer of phages. An important complication of streptococcal disease is *rheumatic fever*. This condition is characterized by fever and inflammations of the small blood vessels. Joint pain is common, but the most significant long-range effect is permanent scarring and distortion of the heart valves, a condition called *rheumatic heart disease*. The damage arises from a reaction of the body's antibodies with streptococcal substances bound to the heart tissue. Antibodies may also react with substances in the heart tissue that are identical to streptococcal substances.

Another manifestation of this reaction may develop in the kidneys, where the condition is called *glomerulonephritis*. Because *S. pyogenes* is commonly present in the human nose and throat, transmission to many other tissues is possible. For example, streptococci may cause disease in open wounds or skin abrasions, resulting in *erysipelas*. If the tissue of the uterus becomes infected during the birth process, *puerperal sepsis*, also known as childbed fever, may develop. Moreover, streptococci may infect the lower respiratory tract and cause streptococcal pneumonia as a complication to viral disease.

Diphtheria

Corynebacterium diphtheriae is a Gram-positive bacillus containing numerous *metachromatic granules* that stain as cytoplasmic dots. The bacteria remain close to one another after multiplying and form a palisade layer, a picket-fence arrangement. The staining and reproductive characteristics are aids to identification. Diphtheria is acquired by the inhalation of respiratory droplets. Infection occurs in the upper respiratory tract near the tonsils. The bacteria produce a potent *exotoxin* that interferes with protein synthesis in epithelial cells. As dead tissue accumulates with mucus, white blood cells, and fibrous material, a leathery *pseudomembrane* forms.Respiratory blockage and death may follow, especially in children. In adults, the toxin often spreads to the bloodstream, where it causes heart damage and destruction of the fatty sheaths surrounding nerves.

Treatment of diphtheria requires both antibiotics for the bacteria and antitoxins to neutralize the toxins. At present, the number of cases of diphtheria in the United States is less than a dozen annually but the disease remains a health problem in many regions of the world, especially parts of Asia. Immunization against diphtheria may be rendered by an injection of diphtheria *toxoid* contained in the diphtheria-pertussis- tetanus (DPT) vaccine. A toxoid consists of toxin molecules treated with formaldehyde or heat to destroy their toxic qualities. It induces the immune system to produce antitoxins that circulate in the bloodstream throughout the person's life.

Pertussis

Pertussis, (*whooping cough*) is one of the more dangerous diseases of childhood years. Although the incidence of this disease has declined substantially from the 250,000 annual cases reported the number of cases in the United States still remains significant. For example, the CDC recorded 3275 cases. Pertussis is caused by *Bordetella* pertussis, a small Gram-negative rod first isolated by Jules Bordet and Octave Gengou commonly known as the Bordet-Gengou bacillus. The bacillus is spread by droplets and uses its pili to adhere to the cilia of the epithelial cells of the upper respiratory tract. There is generally no invasion of tissue beyond this point.

The initial stage of the diseace is marked by general malaise, low-grade fever, and increasingly severe cough. During the next stage disintegrating cells and mucus accumulate in the airways and cause laboured breathing. Victims experience multiple paroxysms of staccato coughs all in one exhalation, followed by a forced inspiration over a partially closed glottis. The rapid inspiration results in the characteristic "whoop". Ten to fifteen paroxysms may occur daily, and exhaustion usually follows each. During the final stage, sporadic coughing may continue for several weeks, even after the bacteria have vanished.

Treatments of pertussis is generally successful when penicillin or erythromycin is administered before the respiratory passageways become blocked. Diagnosis of the disease is performed by obtaining swabs from the posterior pharyngeal wail and identifying *B. pertussis* on selective media. A fluorescent antibody test is also used.

The pertussis component consists of *B. pertussis* cells killed with Merthiolate in the manner first described by Pearl Kendrick and Grace Eldering. Physicians recommend that infants receive the first of these injections at the age of two months, since newborns are highly susceptible to the disease. Use of the pertussis vaccine is not without risk. Report of the American Medical Association estimated that about one in every 300,000 vaccinated individuals may develop brain damage. A few other vaccinees may suffer temporary seizures, and several may experience high fever for a short period of time. Reports such as these have raised public concern about use of the vaccine. The consensus, however, is that many more children have been maimed and killed by pertussis than by the pertussis vaccine.

Meningitis

Meningitis may be caused by viruses, fungi, protozoa or bacteria, and different forms of the disease have different mortality rates. In all forms, the meninges become inflamed, causing pressure on the spinal cord and brain. Victims experience headaches, neckaches, and lower backaches. One dangerous form of meningitis is *meningococcal meningitis*. This disease is caused by *Neisseria meningitidis*, a small Gram-negative encapsulated diplococcus commonly called the *meningococcus*. The diplococci enter the body by droplets, often from a reservoir. In most cases, the disease consists of an influenzalike upper respiratory infection.

However, it sometimes spreads to the bloodstream, where bacterial toxins may overwhelm the body in as little as two hours and cause death. This condition is called *meningococcemia.* In survivors, the diplococci localize on the meninges and cause a characteristic stiff neck and pounding headache. A rash also appears on the skin, beginning as bright red patches, which progress to blue-black spots.

Fifty percent of untreated cases may be fatal. Early diagnosis and treatment of meningococcal meningitis can prevent irreversible nerve damage or death. A principal criterion for diagnosis is the observation of Gram-negative diplococci in samples of spinal fluid obtained by a spinal tap. In active cases of disease, *N. meningitidis* may be cultivated from this fluid.

Treatment with rifampin, penicillin, or sulphonamide drugs is usually recommended. A vaccine containing capsular polysaccharides has been available for immunization. Cases of meningococcal meningitis may be complicated by the formation of lesions in the adrenal glands. This condition, called the *Waterhouse-Friderichsen syndrome,* may be a manifestation of an allergic reaction taking place in the body. Hormone imbalances usually signal its development.

Neisseria meningitidis is a fragile organism that does not survive easily in the environment and must be maintained in nature by person-to-person transfer. Meningococcal meningitis is therefore prevalent where people are in close proximity for long periods of time. Grade school classrooms, military camps, and prisons are examples.

Tuberculosis

Tuberculosis is caused by *Mycobacterium tuberculosis,* the "tubercle" bacillus originally isolated by Robert Koch. It is a small rod that enters the respiratory tract in droplets. Multiple exposures are generally necessary. People who live in urban ghettoes often contract tuberculosis because malnutrition and a generally poor quality of life contribute to the establishment of disease, and overcrowding increases the concentration of bacilli in the air. About 10 percent of people who contract tuberculosis become ill within three months. They experience chronic cough, chest pain, and high fever, and they expel *sputum*, the thick matter accumulating in the lower respiratory tract. Often the sputum is rust-coloured, signalling that blood has entered the lung cavity. The remaining 90 percent of victims exhibit no readily distinguishable symptoms, except malaise, fever, and weight loss. In these cases the body responds to the disease by forming a wall of white blood cells, calcium salts, and fibrous materials around the organisms. As these materials accumulate in the lung, a hard nodule called a *tubercle* arises.

This tubercle may be visible in the chest X ray. However, the bacilli are not killed, and the tubercle may expand as the lung tissue progressively deteriorates. In many instances the tubercle breaks apart and bacteria spread to other organs such as the bone, liver, kidney, and meninges. The disease is now called *miliary tuberculosis* from the Latin milium, for seed.

Tubercle bacilli produce no discernible toxins, but growth is so unrelenting that the tissues are literally consumed, a factor that gave tuberculosis its older name, *consumption*, In children, tuberculosis of the spine is known as *Pott's disease. Mycobacterium tuberculosis* contains a layer of fatty waxy material in its cell wall that greatly enhances resistance to environmental fluctuations. In the laboratory, stain must be accompanied by heat to penetrate this barrier. Once stained, however, the organisms are able to resist decolorization, even when subjected to a 5 percent acid-alcohol solution. Thus the bacilli are said to be acid-resistant, or *acid-fast*. The acid-fast test is an important diagnostic screening tool when used on a sample of patient's sputum.

Physicians treat the disease with such drugs as isoniazid and rifampin, but the regimen must be extended over a minimum of six to nine months without interruption. Early relief, boredom, and forgetfulness often lead the patient to stop taking the medication, and the disease flares up again.

Before the discovery of antituberculosis drugs, patients were sometimes confined to sanataria in the mountains, where fresh air assisted the body's natural healing capacities. Immunization to tuberculosis may be rendered by injections of an attenuated strain of *Mycobacterium bovis*. This species causes tuberculosis in cows as well as humans. The attenuated strain is called *bacille Calmette Guerin*, or *BCG*, after Albert Calmette and Camille Guerin, the two French investigators who developed.

Though the vaccine is utilized in other parts of the world, there is opposition to its use in the United States because some scientists point to the success of early detection and treatment and to the vaccine's occasional side effects. Proponents of BCG, however, stress that 25,000 new cases of tuberculosis occur in the United States annually. They also note the public health estimates that over 12 million people nationwide may carry the bacilli in their bodies and thus may be reservoirs.

Pneumonia

Pneumonia is a microbial disease of the bronchial tubes and lungs. A wide spectrum of organisms, including many different viruses and bacteria, may cause pneumonia. Over 80 percent of bacterial cases are due to

Streptococcus pneumoniae, a Gram-positive encapsulated chain of diplococci traditionally known as the *pneumococcus*. The disease is therefore called *pneumococcal pneumonia*. Pneumococcal pneumonia exists in all age groups, but the mortality rate is highest among older persons and those with underlying medical conditions. *Streptococcus pneumoniae* is usually acquired by droplets or contact and exists in the respiratory tract of the majority of Americans.

However, the natural resistance of the body is high and disease usually will not develop until the defences are compromised. Malnutrition, smoking, viral infections, and treatment with immune-suppressing drugs typify conditions that may precede pneumonia. Patients with pneumococcal pneumonia experience high fever, sharp chest pains, difficult breathing, and rust-coloured sputum. The colour results from blood seeping into the alveolar sacs of the lung as bacteria grow and cause the tissues to deteriorate. Involvement of an entire lobe of the lung is called *lobar pneumonia*. If both lungs are involved, then the condition is called *double pneumonia*. Scattered patches of infection in the respiratory passageways are referred to as *bronchopneumonia*. The term "walking pneumonia" has no clinical significance but is merely a colloquial expression for a mild case of pneumonia.

Traditional drug of choice for pneumococcal pneumonia has been penicillin, with tetracycline and chloramphenicol used for people who are allergic to penicillin. However, public health microbiologists have noted an increasing incidence of resistant strains of *S. pneumoniae*, and in the CDC reported the first strain resistant to all three antibiotics. In such cases erythromycin is useful. Microbiologists have identified over 80 strains of *S. pneumoniae* based on the presence of different components in its capsule. Recovery from one strain therefore tends immunity to that strain alone. The FDA licensed a vaccine for immunization to 23 strains of the organism.

A significant increase in the number of unusual cases of human pneumonia prompted a search for a new infectious agent. At Harvard University, Monroe Eaton isolated a tiny, viruslike agent from the respiratory tracts of patients and cultivated it on media supplemented with blood. The organism was subsequently named *Mycoplasma pneumoniae*, the Eaton agent. The disease came to be known as *primary atypical pneumonia* (PAP), "primary" because it occurred in previously healthy individuals (pneumococcal pneumonia often was a secondary disease), "atypical" because the organism differed from the typical pneumococcus and because symptoms were unlike those in pneumococcal disease. *Mycoplasma pneumoniae* is one of the smallest bacteria causing disease in humans. Mycoplasmas measure about 0.2 μm in size and are *pleomorphic*, that is, they assume a variety of shapes. They have no cell wall and thus no Gram reaction or sensitivity to penicillin.

M. pneumoniae is very fragile and does not survive long periods outside the human or animal host. Therefore it is maintained in nature by passage of droplets from host to host, and multiple exposures are usually necessary to establish disease. PAP resembles viral pneumonia and other respiratory diseases. The patient experiences fever, fatigue, and a characteristic dry, hacking cough. Research indicates that the organisms attach to and destroy the ciliated cells that line the respiratory tract. Blood invasion does not occur and the disease is rarely fatal. Often it is called *walking pneumonia* or *Mycoplasma* pneumonia.

Epidemics are common where crowded conditions exist, such as in college dormitories, military bases, and urban ghettoes. The low humidity and high temperature conditions of summer months appear to favour establishment of disease by drying the respiratory tract and weakening the body's defence to the bacteria. Erythromycin or tetracycline are commonly used as treatments. Antibodies produced against *Mycoplasma pneumoniae* could agglutinate Type O human red blood cells at 4°C but not at 37°C. This observation was developed to the *cold agglutinin screening test (CAST)* in which a patient's serum is combined with red blood cells at cold temperatures to observe clumping. Diagnosis is also assisted by isolation of the organism on blood agar and observation of a distinctive "fried egg" colony appearance.

Carl Friedlander isolated *Klebsiella pneumoniae*, a bacillus that he believed was the sole cause of bacterial pneumonia. In the years thereafter, *Streptococcus pneumoniae* was found to be the major cause, and Friedlander's bacillus was relegated to a minor position. *Klebsiella pneumoniae* is a Gram-negative encapsulated rod responsible for about 5 percent of bacterial pneumonias, The bacillus is acquired by droplets, and often it occurs naturally in the respiratory tracts of humans. Disease appears to be due to factors that compromise the body's natural defences. Examples are alcoholism, chronic emphysema, diabetes, or presence of another disease. *Klebsiella pneumonia* may be a primary disease or a secondary disease. As a primary pneumonia, it is characterized by sudden onset and a gelatinous reddish-brown sputum. The organisms grow over the lung surface and rapidly destroy the tissue, often causing death. Antibiotic therapy is effective if performed vigorously and early. However antibiotic resistance complicates treatment.

Klebsiella pneumonia is a common *nosocomial*, or hospital-acquired, disease. It is spread to hospitalized patients by such routes as clothing, intravenous solutions, foods, and the hands of health-care personnel. Usually, the patient is suffering from an already established disease, hence the "secondary" connotation. The symptoms, while similar to those in the primary disease, are compounded by the patient's other illness and the weakened condition of the body. *K. pneumoniae* also causes urinary tract and gastrointestinal infections.

Microbiologists considered *Serratia marcescens* a nonpathogenic bacillus and often used it as a test organism. Today they view this Gram-negative rod as a widespread cause of nosocomial disease. A patient may be predisposed to infection by such conditions as chronic illness, impaired immunity due to immunosuppressive therapy, radiation, or surgical treatments such as urinary catheterization, lumbar puncture, or blood transfusion. *Serratia pneumonia* is accompanied by patches of bronchopneumonia and, in some cases, substantial tissue destruction in the lungs. The major clinical problem in treatment is resistance to antibiotic therapy, apparently due to R factors. *Serratia marcescens* as a cause of eye infections, bone disease, arthritis, and meningitis. In addition, it is a widespread agent of urinary tract diseases.

Q fever

The term *Q fever* was first used by E.H. Derrick to describe an illness that broke out among workers at a meat-packing plant in Australia. The "Q" is thought to be derived either from "query," meaning unknown, or from Queensland, the province in which the disease occurred. The rickettsia that causes Q fever is named *Coxiella burnetii*, after H.R. Cox, who isolated it in Montana in 1935, and Frank Macfarlane Burnet, who studied its properties. Q fever is prevalert worldwide among livestock, especially dairy cows, goats, and sheep, and outbreaks may occur wherever these animals are raised, housed, or transported.

Transmission among livestock and to humans is accomplished primarily by airborne dust particles as well as by respiratory droplets and ticks. In addition, humans may acquire the disease by consuming raw milk infected with *C. burnetii* or milk that has been improperly pasteurized. Patients with Q fever experience severe headache, high fever, a dry cough, and occasionally, lesions on the lung surface. The mortality rate is low, and treatment with tetracycline is effective. A vaccine is available for high-risk individuals. Q fever is an occupational hazard of those who work with domestic animals. In both instances, investigators found *C. burnetii* in sheep and goats. An outbreak was reported among California slaughterhouse workers. The organism is among the most heat-resistant of all bacteria, and its elimination is a prime object of milk pasteurization.

Foodborne and waterborne bacterial diseases

Typhoid fever

Typhoid fever drew the attention of microbiologists a century ago and was studied by Karl Eberth and Georg Gaffky, both associates of Robert Koch. Eberth observed the causative organism, Salmonella typhi, in spleen tissue and Gaffky isolated it in pure culture four years later. *Salmonella typhi*, a Gram-negative rod, displays high resistance to environmental conditions outside the body. This factor enhances its ability to remain alive for long periods of time in water, sewage, and certain foods. *S. typhi* causes disease only in humans and is transmitted by the five Fs: flies, food, fingers, feces, and formites. *S. typhi* is acid-resistant, and with the buffering effect of food and beverages, it generally survives passage through the stomach. In the small intestine it invades the tissues, causing deep ulcers and bloody stools, but little diarrhea.

Blood invasion follows, and after a few days, the patient experiences mounting fever, lethargy, and delirium. The abdomen becomes covered with *rose spots*, an indication that blood is hemorrhaging in the skin. Bowel perforation and gall bladder infection are other possible complications.

Treatment of typhoid fever is generally successful with the antibiotic chloramphenicol. In the last decade, however, antibiotic resistance due to R factors has been noted, and other drugs have been substituted. About 5 percent of recoverers continue to harbor and shed the organisms for one year or more. Individuals such as these are termed *carriers*. Because they are possible sources of disease to others, the public health department usually monitors the activities of carriers until they are found free of bacteria. On occasion, gall bladder removal may be recommended since this is often the site in which the bacilli remain. Immunization to typhoid fever may be rendered by injecting a vaccine consisting of dead *Salmonella* cells. Enthusiasm for this procedure is mixed, however, because individuals may still contract the disease and because wide-scale epidemics are rare in the United States, about 500 cases are reported annually to the CDC.

Salmonellosis

Salmonellosis is caused by hundreds of serotypes of *Salmonella*. Serotypes are used for *Salmonella* because of uncertain relationships existing among the organisms. The most common causes of salmonellosis include *S. typhimurium*, *S. heidelberg*, *S.enteritidis*, and *S. newport*. These organisms are Gram-negative rods with peritrichous flagella that usually enter the human intestine in foods. After an incubation period of one to three days, the patient with salmonellosis experiences fever, nausea, vomiting, diarrhea, and abdominal cramps.

Ulceration of the intestine is usually mild compared to that in typhoid fever, and blood invasion is uncommon. The symptoms may last a week or more and generally, the disease is allowed to run its course rather than treat the patient with antibiotics that may damage the tissues. Dehydration may occur in some patients and fluid replacement may be necessary.

Diagnosis usually consists of isolating the *Salmonella* serotype from stool specimens or rectal swabs. Often the infection is called *gastroenteritis* or simply, *enteritis*. With increased awareness and modern methods of detection, salmonellosis has been linked to a broad variety of foods.

Shigellosis (bacterial dysentery)

Shigella bacilli are small gramnegative rods found mainly in humans and other primates. They cause digestive disturbances ranging from mild diarrhea to a severe and sometimes fatal dysentery. *Dysentery* is a syndrome manifested by waves of intense abdominal cramps and frequent passage of small-volume, bloody mucoid stools. For many years, *Shigella* diseases were known as *bacterial dysentery*. However, watery diarrhea without blood or mucus is a more common symptom than dysentery, and the name shigellosis is currently preferred. *Shigellosis* may be caused by any of four serotypes of *Shigella,* including *Sh. sonnei, Sh. dysenteriae, Sh. flexneri,* and *Sh. boydii*. Of these, *Sh. sonnei* is the most common.

Humans can ingest the organisms in contaminated water as well as in many foods, especially eggs, vegetables, shellfish, and dairy products. Shigella usually penetrates the epithelial cells lining the intestine and, after two to three days, produces sufficient enterotoxins to encourage the cells to release water. Infection of the small intestine produces watery diarrhea, but infection of the large intestine results in the expulsion of bloody mucoid stool's characteristic of dysentery. Identification of the causative agent is confirmed by the isolation of *Shigella* from stool specimens. Most cases of shigellosis subside within one week without complications.

However, patients who lose excessive fluids must be given salt tablets or intravenous injections of salt solutions. This treatment is most often applied to infants and debilitated adults in whom circulatory collapse may lead to death. Antibiotics are sometimes effective, but many strains of *Shigella* are resistant due to R factors, and the shigellosis may be fatal. Recoverers generally become carriers for a month or more, and continue to shed the bacilli in their feces.

Cholera

Cholera is caused by *Vibrio cholerae,* a curved Gram-negative rod first isolated by Robert Koch. The bacilli enter the intestinal tract in contaminated water or food. Vegetables from fields fertilized with nightsoil are a possible source. The effluent from aircraft has also been implicated. Another possibility is shellfish such as crabs, as shown by a Louisiana. *Vibrio cholerae* is extremely susceptible to stomach acid. However, if sufficient numbers are ingested, enough will remain to colonize the intestine. As the bacilli move along the intestinal epithelium, they secrete an enterotoxin that stimulates the unrelenting loss of fluid characteristic of the disease. Antibiotics such as tetracycline may be used to kill the bacteria, but the key treatment is replacement of lost fluids and restoration of the water balance in the body. Often, this entails intravenous injections of salt solutions and the consumption of glucose and salts.

The current pandernic began in 1961 in Indonesia and now involves about 35 countries. Travelers to cholera regions of the world are immunized with preparations of dead bacteria and receive protection for approximately six months. The most important preventive measures, however, are sanitation, personal hygiene, and care in food preparation. Outbreaks in the United States arc rare, but the disease is estimated to cause an annual four to five million deaths throughout the world.

Microbiology and public health

Microorganisms contribute to public health in a number of ways. For example, they are used to produce a broad assortment of foods and dairy products that make up a regular part of our diet. In addition, microorganisms play a dominant role in numerous cycles of elements in the environment and forge a link between what is useless and what is useful to other living things.

Food spoilage

Food is considered spoiled when it has been altered from the expected norm. Usually, the food has an unpleasant appearance, aroma, and taste. Sometimes, however, these signs may be difficult to detect such as when staphylococci deposit enterotoxins in food or when too few bacteria grow to cause a perceptible change. Consumption of toxins or microorganisms may cause a number of food poisonings or infections. Contaminating microorganisms enter foods from a variety of sources. Airborne organisms, for example, fall onto fruits and vegetables, then penetrate the product through an abrasion of the skin or rind. Crops carry soilborne bacteria to the processing plant. Shellfish concentrate organisms by straining contaminated water and catching the organisms in their filtering apparatus. Rodents or arthropods transport microorganisms on their feet and body parts as they move about among foods. Human handling of foods also provides a source of contamination.

For example, bacteria from an animal's intestine contaminate meat handled carelessly by a butcher. Of even more concern are raw vegetables such as those obtained at salad bars. For instance, 107 Maryland residents contracted shigellosis after eating from the salad bar in the cafeteria at a military hospital. Earlier that year, 123 cases of hepatitis A were diagnosed in patients who dined at a salad bar restaurant in Lubbock, Texas. In both cases, investigators believed that food handlers were the source of pathogenic microorganisms.

Food is basically a culture medium for microorganisms, soo that the chemical and physical properties of food have a significant bearing on the type of microorganisms growing on or in it. Water, for instance, is a prerequisite for life, and therefore the food must be Moist, with a minimum water content of 18 to 20 percent. Microorganisms do not grow in foods such as beans, rice, and flour because of their low water content. Another important factor is pH. Most foods fall into the slightly acidic range on the pH scale, and numerous species of bacteria multiply under these conditions. In foods with a pH of 5.0 or below, acid-loving molds often replace the bacteria. A third property of the food is its physical structure. A steak, for example is not likely to spoil quickly because microorganisms cannot penetrate the meat easily. However, an uncooked hamburger rapidly deteriorates since microorganisms exist within the loosely packed ground meat as well as on the surface. Similar conditions exist for bologna, frankfurters, and other processed meats.

Chemical composition of the food may be another determining factor in the type of spoilage encountered. Fruits, for instance, support organisms that metabolize carbohydrates, whereas meats attract protein decomposers. Starch-utilizing bacteria and molds are often found on potatoes, corn, and rice products. The presence of certain vitamins encourages particular microorganisms to proliferate, while the absence of other vitamins provides natural resistance to decay. Oxygen and temperature are other considerations. Vacuum-sealed cans of food do not support the growth of aerobic bacteria, nor do vegetables or most bakery products support anaerobes. Similarly, the refrigerator is usually too cold for the growth of human pathogens but the warm hold of a ship or a humid, hot warehouse storeroom is an environment conducive to the growth of these pathogens.

It is common knowledge that contamination is more likely in cooked food at warm temperatures than in refrigerated cooked food. The food industry recognizes three groups of foods loosely defined on the basis of their chemical and physical properties. The highly perishable foods are those that spoil rapidly. They include poultry, eggs, meats, most vegetables and fruits, and dairy products. Other foods such as nutmeats, potatoes, and some apples are considered semiperishable, because they spoil less quickly. The nonperishable foods are often stored in the kitchen pantry. Included in this group are cereals, rice, dried beans, macaroni products, flour, and sugar.

Food spoilage often ocuuring due to the naturally occurring chemistry of contaminating microorganisms. Yeasts, for instance, live in apple juice and convert the carbohydrate into ethyl alcohol, a product that gives spoiled juice an alcoholic taste. Certain bacteria convert food proteins into amino acids, then break down the amino acids into foul-smelling end-products. Cysteine digestion, for example, yields hydrogen sulfide, which imparts a rotten egg smell to food. Tryptophan digestion yields indole and skatole, which give food a fecal odor. Two other possible products of the microbial metabolism of carbohydrates are acid, which causes foods to become sour, and gas, which causes scaled cans to swell.

Moreover, when fats break down to fatty acids as in spoiled butter, a rancid odor or taste may evolve. Capsule production by bacteria causes food to become slimy, and pigment production imparts color. In numerous historical incidents, the red pigment from Serratia marcescens in bread has been interpreted as sign of blood. Some foods resist spoilage naturally because they contain antimicrobial chemicals. For example, the white of all egg has lysozynic, an enzyme that digests the cell wall of Gram-positive bacteria. Garlic contains certain compounds that inhibit many bacteria. Oil of cloves appears to have heating tendencies, while radish and onion extracts both retard bacterial growth.

Fish and meats are originally free of contamination because the muscle tissues of living animals are normally sterile. Spoilage organisms enter during handling, processing, packaging, and storage. For example, if a piece of meat is ground for hamburger, microorganisms from the surface accumulate in the teeth of the grinder along with other dust borne organisms.

Bacteria from preparers hands or from a sneeze compound the problem. The grinder may be cleaned well, but rarely is it sterilized. In some meat markets, butchers refrigerate the head of the grinder until the next use to control contamination. Processed meats such as luncheon meats, sausages, and frankfurters represent special hazards because they are handled often. Also, natural sausage and salami casings made from animal intestines may contain residual bacteria, especially botulism spores. When preparers pack such casings tightly with meat, Clostridium butulinum may multiply and produce its powerful toxins. People recognized the symptoms of "sausage poisoning" and coined the name botulism, which translates loosely to sausage. Organ meats such as livers, kidneys and sweetbreads are less compact tissue than muscle and thus spoil more quickly. Moreover, because these organs contain many natural filtering tissues, bacteria tend to be trapped here.

Extent of contamination in meats often consists of a harmless greening" seen on the surface of a steak. This discoloration is commonly due to the Gram-positive rod *Lactobacillus,* or the Gram-positive coccus *Leuconostoc.* The green color results from pigment alteration in the meat and represents no hazard to the consumer. Slime formation on the outer casings and souring in frankfurters, bologna, or processed meats may also result from these organisms as well as from *Streptococcus* species. Microbiologists can often trace spoilage in *fish* to the water from which it is taken or held. Fish tissues deteriorate rapidly and the filleting of fish on a blood-stained block encourages contamination. Shellfish are of particular concern because they commonly obtain their food by filtering particles from the water. Clams, oysters, and mussels therefore concentrate such pathogens as hepatitis A viruses, typhoid bacilli, cholera vibrios, or amebic cysts.

Spoilage in poultry and eggs may reflect human contamination, but it usually stems from microorganisms that have infected the bird. Members of the genus *Salmonella,* for example, cause diseases in *chickens and turkeys,* then pass to consumers via poultry and egg products. New York was traced to contaminated eggs used to make cheese lasagna and stuffed shells. Processed foods such as chicken pot pies, whole egg custard, mayonnaise, egg nog, and egg salad may also be sources of salmonellosis. Psittacosis is another problem because *Chlamydia psittaci* infects poultry. For instance, 27 employees in a turkey processing plant contracted psittacosis while preparing consumer products from diseased turkeys.

Spoilage organisms can also enter poullry products from the birds' visceral organs during improper handling, or from water used for cleaning. Eggs are normally sterile when laid, but the outer waxy membrane as well as the shell and inner shell membrane may be penetrated by bacteria after several hours. *Proteus* species cause black rot in eggs as hydrogen sulfide gas accumulates from the breakdown of cysteine. This gives eggs a rotten odor. Other spoilages in eggs include green rot from *Pseudomonas* species and red rot from the growth of *Serrafia marcescens*. Green rot results from a fluorescent pigment that causes the egg to glow when placed in front of an ultraviolet light. Red rot gives the yolk a blood-red appearance.

Primary focus of contamination in an egg is the yolk rat er than the white. This is due to the nutritious quality of the yolk and because the white has a pH of approximately 9.0. Also the lysozyme in egg white is inhibitory to Gram-positive bacteria. It should be noted that even though eggs are hard-boiled, contamination is not always prevented.

In the production of bakery products, ingredients such as flour, eggs, and sugar are generally the sources of spoilage organisms. Although most contaminants are killed during baking, some bacterial and mold spores survive because bread is heated internally to 100°C for only about nine minutes. Members of the spore forming genus Bacillus commonly survive, and as they proliferate, their capsular material accumulates, giving the bread a soft cheesy texture with long, stringy threads. The bread is said to be ropy. Cream fillings and toppings in bakery products provide excellent chemical and physical conditions for bacterial growth. For example, custards made with whole eggs may be contaminated with Salmonella species, and whipped cream may contain dairy organisms such as species of Lactobacillus and Streptococcus. The acid produced by these bacteria results in a sour taste. High sugar environments of chocolate toppings and sweet icings support the growth of fungi. Most bakery products should be refrigerated during warm summer months.

In the case of grains, two types of spoilage are seem to be important in public health microbiology. The first type of spoilage is caused by the ascomycete *Aspergillus flavus*. This mold produces *aflatoxins,* a series of toxins that accumulate in stored grains such as wheat, peanuts, soybeans, and corn. Scientists have implicated aflatoxins in liver and colon cancers in humans, The toxins are consumed in grain products as well as meat from animals that feed on contaminated grain. The second type of grain spoilage is caused by *Claviceps purpurea,* the cause of *ergot poisoning* (ergotism). Rye plants are particularly susceptible to this type of spoilage but wheat and barley grains may also be affected. The toxins deposited by C. purpurea may induce convulsions and hallucinations when consumed.

Experience taught humans they could overcome these difficult times by preserving foods. Among the earliest methods was drying vegetables and strips of meat and fish in the sun. Foods could also be preserved by salting, smoking, and fermenting. Individuals could now trek far from their native habitat, and soon they took to the sea and moved overland to explore new lands. Lazaro Spallanzani attempted to disprove spontaneous generation by showing that beef broth would remain unspoiled after being subjected to heat. Nicholas Appert applied this principle to a variety of foods. Neither Appert nor his contemporaries were quite sure why food was being preserved, but it was clear that the spoilage could be retarded by prolonged heating. The significance of microorganisms as agents of spoilage awaited Pasteur's classic experiments with Wine several generations later.

Modern preservation methods arc mere extensions of these principles. Though today's methods are sophisticated and technologically dynamic, advances in preservation processes are counterbalanced by the great volumes of food that must be preserved and the complexity of food products. Thus the problems that early humans faced do not differ fundamentally from those confronting modern food technologists.

Preventing foodborne disease

People are then affected by the organism or the toxin it has produced in the food. In the former case, a food infection is established; in the latter, a food poisoning or intoxication occurs. Food infections are typified by typhoid fever, salmonellosis, cholera, and shigellosis, all of which are of bacterial origin. Among the protozoan infections, amebiasis, balantidiasis, and giardiasis represent foodborne diseases. Viral infections are exemplified by hepatitis A. Food intoxications include botulism, staphylococcal food poisoning, and ctostridial food poisoning. Since full discussions of these diseases are presented elsewhere in this text, we shall not pause to examine them individually.

Public health microbiologists estimate that between two and ten million people are affected by foodborne disease annually. Many episodes require medical attention, but the vast majority of patients recover rapidly without serious complications. In many cases the incident might have been avoided by taking some basic precautions. For example, unrefrigerated foods are a prime source of staphylococci and Salmonella serotypes, and perishable groceries such as meats and dairy products should not be allowed to warm up while other errands are performed.

The hands should always be washed thoroughly after handling raw vegetables or salad fixings to avoid cross-contamination of other foods. It is well to cook meat from a frozen or partly frozen state; it this is impossible, the meat should be thawed in the refrigerator. Cutting-I boards should be cleaned with hot, soapy water after use, and old cutting boards with cracks and pits should be discarded.

Leftovers are implicated in most outbreaks of foodborne disease. It is therefore important to refrigerate leftovers promptly and keep them no more than a few days. Thorough reheating of leftovers, preferably to boiling, also reduces the possibility of illness. Many instances of foodborne disease occur during summer months when foods are taken on picnics where they cannot be refrigerated. In general, dairy foods such as custards, cream pies, pastries, and salads should be excluded from the picnic menu. For outdoor barbecues, one dish should be used for carrying hamburgers to the grill and another dish for serving them. Many of these principles apply equally well to fall and winter tailgate parties.

Foods suspected of contamination should not be tasted to confirm the suspicion, but should be discarded immediately. If doubtful, it is well to boil the food for a minimum of 10 minutes and thoroughly wash the utensil used to stir the food. Bulging or leaking cans must be discarded in a way that will not endanger other people or animals. One final point should be made: there is no such thing as ptomaine poisoning. Ptomaines are a group of nitrogen-containing chemical compounds that have a strong odor. At one time it was believed that ptomaines were a cause of foodborne disease, but the compounds have been fully exonerated.

Foods from microorganism

Social customs and traditions have brought acceptance of a variety of foods produced by microorganisms. Some individuals regard these foods as "spoiled," but to many people, the food is fermented." Fermented foods have three things in common: they are less vulnerable to extensive spoilage than unfermented foods; they are less likely to be vectors of' foodborne illness than unfermented foods; and they have not only been accepted by the cultures in which they were developed, but in sonic cases, they are considered delicacies. It is conceivable that ancient peoples were first attracted to the preservative qualities of fermented foods and coincidentally learned to appreciate their tastes.

Sauerkraut

Sauerkraut (German for "sour cabbage") is not only a well-preserved and tasty form of cabbage but also is nutritionally sound. For instance, the vitamin C content of sauerkraut is equivalent to that of citrus fruits, and sauerkraut was often taken on British sea voyages to prevent scurvy because citrus fruits were too expensive. *Sauerkraut* is prepared commercially by adding salt to shredded cabbage and packing the cabbage tightly to stimulate anaerobic conditions.

The first organisms to multiply are species of *Leuconostoc,* a Gram-positive coccus found naturally in the cabbage. These bacteria ferment carbohydrates in the plant cells and produce acetic and lactic acids. After some days, the acids lower the pH of the cabbage to about 3.5. Species of *Lactobacillus* then proliferate, and the additional lactic acid they produce by fermentation further reduces the pH to about 2.0. The salt helps retard mold contamination while drawing juices out of the plant cells. A compound called diacetyl (the flavoring agent in butter) is produced by *Leticonostoc,* adding aroma and flavor.

Pickles

Pickle is practically synonymous with "pickled cucumber." Over 37 types of dill, sour, and sweet *pickles* have been categorizes but essentially the fermentations are similar. Cucumbers are placed in a salt solution of 8 percent or higher, at which point the cucumber changes color from bright green to olive green. Next comes curing. Three groups of microorganisms arc important to the fermentation and curing of cucumbers. *Enterobacter aerogenes,* a Gram-negative rod, produces large amounts of

CO_2, which takes up all the air space and establishes anaerobic conditions. *Lactobacillus* and *Leuconostoc* species then dominate and form abundant amounts of acid that softens the tissues and sours the cucumbers. Finally, certain yeasts grow and establish many flavors associated with ripe pickles. Dill, garlic, and other herbs and spices are added to finish the product.

Vinegar

Vinegar is a fermented food that traditionally has been made by tile spontaneous souring of wine. Indeed, the word "vinegar" is derived from the French vinaigre, which means sour wine. Yeasts ferment the fruit juice until the alcohol concentration is about 10 to 20 percent. Machines then spray the juice into a tank containing the bacterium *Acetobacter aceti* growing on the surface of wood shavings, gravel, or other substrate. As alcohol percolates through, bacterial enzymes convert it to acetaldehyde and then acetic acid. The vinegar recirculates several times before collection at the bottom of the tank. Residual alcohol evaporates in the heat, and the product usually has an acetic acid content of about 3 to 5 percent. The flavor of vinegar is determined by oils, sugars, and other compounds produced by the bacteria plus the residue of organic compounds in the wine from which it was made.

Several other fermented foods are also worthy of note. One *example, soy sauce*, is made from roasted soybeans and wheat inoculated with the fungus *Aspergillus oryzae* and allowed to stand for three days. The fungal-covered product, called koji, is then added to a solution of salt and microorganisms, and aged for about a year. During this time, lactobacilli produce acid, and yeasts produce small amounts of alcohol. Together with the fungal products, the acid and alcohol determine the flavor.

Another example is fermented sausages. These are generally produced as dry or semidry products and include pepperoni from Italy, thuringer from Germany, and polsa from Sweden. Curing and seasoning agents are first added to ground meats, followed by stuffing into casings and incubation at warm temperatures. Mixed acids produced from carbohydrates in the meat give the sausage its unique flavour and aroma. Cocoa and coffee also owe their flavor in part to microorganisms. Cocoa derives some of its taste from the microbial fermentation that helps remove cocoa beans from the pulp covering them in the pod. Likewise, coffee is believed to obtain some of its flavor from the fermentation of coffee berries when the beans are soaked in water to loosen the berry skins before roasting.

Influence of microorganisms on animal nutrition is evident in the animal food called silage. Silage is made in huge, cylindrical silos that commonly stand adjacent to barns. The farmer packs the silo with corn, grain stalks, grass, potatoes, and virtually anything that can be fermented. During storage, innumerable bacteria ferment the plant carbohydrates in a process similar to that for sauerkraut and pickles. Fermentation yields a broad mixture of acids, aldehydes, and ketones as well as diacetyl, which is relished by cattle. After several weeks, the well-preserved plant material is sweet smelling, succulent, and nutritious to animals.

Chapter 3

Virology

Viruses are recognized as particles of nucleic acid and protein, often with a covering membrane. They replicate in living cells and cause a number of important diseases such as genital herpes, influenza, and infectious mononucleosis.

Viruses vary considerably in size, shape, and chemical composition, and the methods used in their cultivation and detection are completely different than for other microorganisms. The term 11 rabies virus" is still a common expression in microbiology, but with a vastly different meaning than in Pasteur's time. No one person discovered the viruses. Instead, an understanding of the viruses evolved with the general understanding of the germ theory of disease.

Different diseases had recognizable patterns and, although a bacterium, protozoan fungus or other agent could be isolated for most diseases, some had no identifiable agent. Many of these diseases would become the viral diseases. A mechanical device, the filter, led to the early interest in viruses. Charles Chamberland, an associate of Pasteur, devised a porcelain filter to trap the smallest known bacteria, and the Russian pathologist Dimitri A. Iwanowski, used Chamberland's filter in his studies on tobacco mosaic disease.

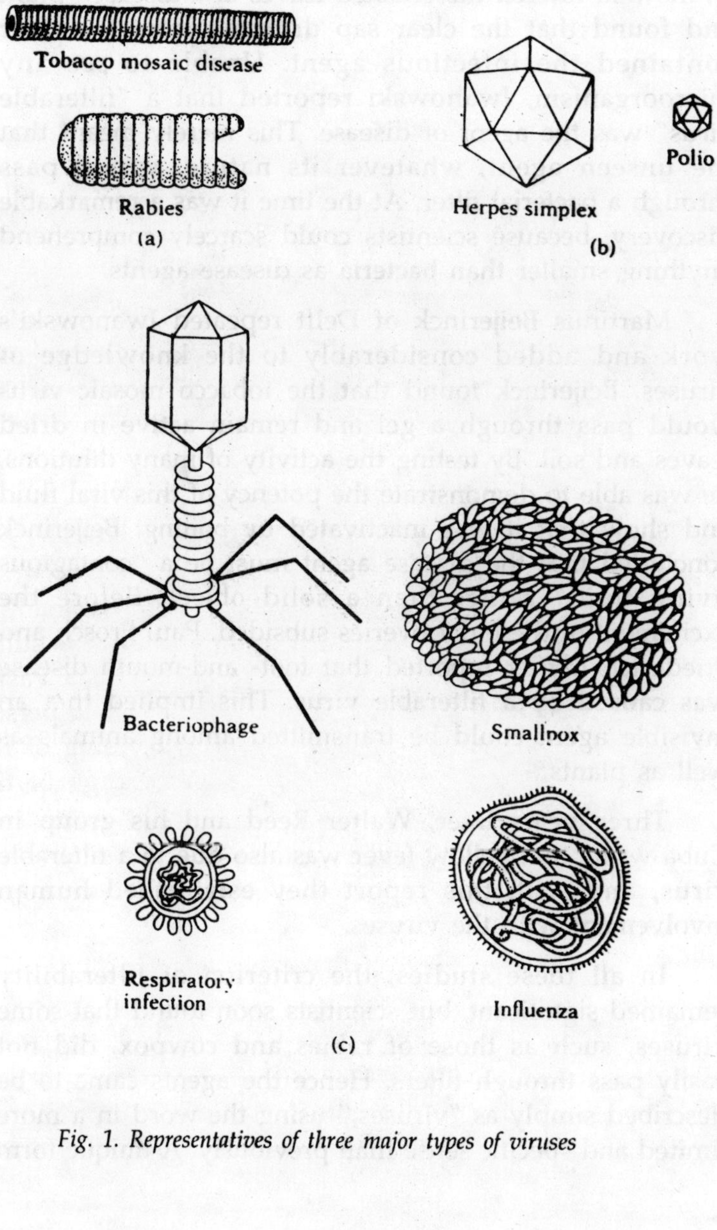

Fig. 1. Representatives of three major types of viruses

In tobacco mosaic disease, the tobacco leaves shrivel and assume a mosaic appearance before dying. Iwanowski filtered the crushed leaves of a diseased plant and found that the clear sap dripping from the filter contained the infectious agent. Unable to see any microorganism, Iwanowski reported that a "filterable virus" was the agent of disease. This merely meant that the unseen agent, whatever its nature, would pass through a bacterial filter. At the time it was a remarkable discovery, because scientists could scarcely comprehend anything smaller than bacteria as disease agents.

Martinus Beijerinck of Delft repeated Iwanowski's work and added considerably to the knowledge of viruses. Beijerinck found that the tobacco mosaic virus would pass through a gel and remain active in dried leaves and soil. By testing the activity of many dilutions, he was able to demonstrate the potency of this viral fluid and show that it was inactivated by boiling. Beijerinck concluded that the disease agent must be a "contagious living fluid" rather than a solid object. Before the excitement of these discoveries subsided, Paul Frosch and Friederich Loffler reported that foot- and-mouth disease was caused by a filterable virus. This implied that an invisible agent could be transmitted among animals as well as plants.

Three years later, Walter Reed and his group in Cuba wrote that yellow fever was also due to a filterable virus, and with this report they established human involvement with the viruses.

In all these studies, the criterion of filterability remained significant, but scientists soon found that some viruses, such as those of rabies and cowpox, did not easily pass through filters. Hence the agents came to be described simply as "viruses," using the word in a more limited and specific sense than previously. A unique form of virus was discovered by the English bacteriologist Frederick Twort, and independently by the French scientist Felix d'Herrelle of the Pasteur Institute. The virus of Twort and d'Herrelle came to be called the *bacteriophage* ("bacteria-eater") for its ability to destroy bacteria. This observation stemmed from experiments in which a drop containing virus was placed in a broth culture of bacteria; the bacteria dissolved in minutes.

D'Herrelle believed that bacteriophages (or simply, phages) could be used to kill bacteria in the body, but investigators later found that phages were easily eliminated from the body and were highly specific for the bacterial strain they attacked. Bacteriophages have since become important tools in transduction research and in bacterial identification.

When the germ theory was established, some scientists believed that cancer might prove to be a bacterial disease. But no bacterium could be located. A relationship between viruses and cancer was suggested by Wilhelm Ellerman and Olaf Bang. Their investigations showed that leukemia in chickens could be induced by a preparation of virus. Another researcher, Francis Peyton Rous, demonstrated that cancer of the connective tissues could be traced to a virus.

Viruses were a type of invisible microorganism below the resolving power of available microscopes. As the years passed, a long list of viral diseases had developed, and work with plant viruses had been productive because of the ease of viral cultivation. For example, virologists learned that tobacco mosaic viruses were composed exclusively of nucleic acid and protein. Work with animal viruses, however, was less productive because cultivation in live animals was a difficult procedure.

When a living analog of Koch's nutrient agar was discovered for viruses. In that year Alice M. Woodruff and Ernest W. Goodpasture of Vanderbilt University published a paper describing the use of fertilized chicken eggs as a nutrient for the cultivation of some viruses. The shell of the egg served as a type of natural Petri dish for the nutrient medium, and viruses multiplied within the Chick embryo. The assumption had been that viruses, though incredibly small, were living things, but a discovery by Wendel M. Stanley of Rockefeller Institute raised doubts. Three protein enzymes of the human body were crystallized by biochemists, and Stanley made the startling announcement that he had crystallized tobacco mosaic viruses. Crystals are composed of identical molecules, and some scientists now suggested that viruses were as lifeless as any crystalline chemical molecules.

Virologists pointed out, however, that the viruses could replicate, cause fevers, and elicit antibody responses, properties not associated with chemical molecules. Stanley's unprecedented discovery was rewarded with the Nobel Prize in Chemistry. It opened a debate on the living or nonliving nature of viruses that is still unsettled, and it cut across preconceived ideas that only living things could cause disease. Researchers in the brought virology into the modern era. Two developments of that period were particularly significant. One was the electron microscope, which allowed viruses to be visualized and hence, their functions to be understood. The other was the test tube tissue cultivation of viruses, which made vaccines possible for the control of some viral diseases.

The filter led to the early notion of viruses, so too, another technological device, the electron microscope, revealed the nature of the viruses. The innovative feature of the electron microscope is the incredibly short wavelength of the electron beam, which dramatically increases the resolving power of the instrument. To understand the principle involved, consider the difference between running and walking on a beach. With a running stride, a person misses most grains of sand,·just as visible light with its long wavelength passes over the viruses without touching them, When the stride is shortened, as in walking, more grains of sand are contacted just as the electron beam with its short wavelength contacts more viruses.

Polio is a disease that attacks people indiscriminately and often leads to permanent paralysis. Attempts at vaccine production were stymied by the inability to cultivate polio viruses outside the animal body, but John Enders, Thomas Weller, and Frederick Robbins of the Children's Hospital in Boston solved that problem. Meticulously, they developed a test tube Medium of nutrients, salts, and pH buffers in which living cells would remain alive. They used the newly, developed antibiotics to retard bacterial growth, and changed the fluid often to remove bacterial waste products. In the living cells, polio viruses replicated to huge numbers, and by the Salk and Sabin vaccines became reality. Enders, Weller, and Robbins did not invent tissue cultivation of viruses but their work showed it had a practical value for vaccine production. They shared the Nobel Prize in Physiology or Medicine for their work. The electron microscope and the test tube cultivation of' viruses paved the way for the discoveries in virology that have come forth in our era.

Structure of viruses

Viruses are the smallest agents able to cause disease in living things. They range in size from the 250 nanometers (nm) of poxviruses to the 20 nm of parvoviruses. At the upper end of the spectrum, the viruses approximate the chlamydiae and mycoplasmas in size; at the lower end, they are about as small as the diameter of a DNA molecule. Viruses may appear in several shapes. Certain viruses, such as rabies and tobacco viruses, exist in the form of a *helix* and are said to have helical symmetry. The helix is a tightly wound coil resembling a corkscrew or spring.

Other viruses, such as herpes simplex and polio viruses, have the shape of an *icosahedron* and hence, icosahedral symmetry. The icosahedron is a polyhedron with 20 triangular faces and 12 corners. Certain viruses have a combination of helical and icosahedral symmetry, a construction described as complex. Poxviruses, for example, are brick-shaped with filaments or tubes wrapped in a swirling pattern around the virus. Certain bacteriophages also have complex symmetry, with an icosahedral head and a collar and tail assembly in the shape of a helical sheath.

Viruses consist of two basic components: a core of nucleic acid called the *genome,* and a surrounding coat of protein known as the *capsid*. The genome contains either DNA or RNA, but not both, and the nucleic acid occurs in double-stranded or single-stranded form.

Usually the nucleic acid is unbroken, but in some instances it exists in segments. The genome is usually folded and condensed in icosahedral viruses, and coiled in a helical fashion in helical viruses. The capsid gives shape to the virus and is responsible for the helical, icosahedral, or complex symmetry. Generally, the capsid is subdivided into individual protein subunits called capsomeres. The number of capsomeres is characteristic for a particular virus. For example, 162 capsomeres exist in the viruses of genital herpes and the 252 capsomeres are located in adenoviruses, which cause the common cold. The capsid provides a protective covering for the genome because the chemical construction of capsid amino acids resists temperature, pH, and other environmental fluctuations. In some viruses, capsid proteins are organized into enzymes to assist cell penetration during the replication process. Also, the capsid is the structure that stimulates the immune system during periods of disease. The capsid plus the genome is called the *nucleocapsid*.

Many viruses are surrounded by a flexible membrane known as an *envelope*. The envelope is composed of lipids and protein and is essentially similar to the host cell membrane, except that it contains viral-specified components. It is acquired from the host cell during the replication process and is unique to each type of virus. In some viruses, such as influenza and measles viruses, the envelope contains functional projections known as spikes. The spikes often contain enzymes to assist the attachment of viruses to host cells. Enveloped viruses may lose their infectivity when the envelope is destroyed.

Compared to a prokaryote such as a bacterium, a virion is extraordinarily simple. As we have seen, it consists essentially of a fragment of nucleic acid, a protein coat, and in some cases, an envelope. Occasionally enzymes are found in the capsid or envelope, but these instances are the exception. Virions lack the chemical machinery for generating energy and synthesizing large molecules. Therefore they must rely upon the structures and chemical components of their host cells to replicate themselves.

Replication of bacteriophages

Process of viral replication is among the most remarkable events of science. A virion invades a living cell a thousand or more times its size, utilizes the metabolism of the cell, and produces copies of itself while destroying the cell. The virion cannot replicate independently, but within the cell, the replication takes place with high efficiency. Replication has been studied in a wide range of virions and their host cells. Among the best known processes of replication is that carried on by bacteriophages of the T-even group (T for "type").

Bacteriophages T2, T4, and T6 are in this group. They are large, complex DNA virions with the characteristic head and tail of bacteriophages but without an envelope. We shall use their replication cycle in *Escherichia coli* as a model for the viruses. The first step in the phage replication is the *union* of phage and host bacterium. Some specificity is involved in this union because adsorption sites on the phage tail fibers must match a receptor site on the bacterial cell wall. For some other phages the receptor sites are on bacterial flagella or pili. Next the phage *penetrates* the host cell. This process begins when enzymes in the phage tail degrade a portion of the bacterial cell wall. The tail sheath then contracts, and the tail core drives through the cell wall and cell membrane. DNA from the phage head now enters the cytoplasm while the capsid remains outside.

Viral DNA transcribes mRNA, and the bacterial cell produces enzymes for phage synthesis. Some enzymes degrade cell components to be used for viral parts, while other enzymes synthesize viral parts. Many of the host cell's ribosomes, enzymes, and amino acids are utilized. Because viral proteins are repeating units of capsomeres, a simple genetic code may be repeated many times over. AIM several minutes, complete phages appear in the cytoplasm. The assembly proceeds in a step-by-step sequence and is complete in about 30 minutes.

The last stage is the release stage, often called lysis. The enzyme lysozyme coded by a phage gene, degrades the bacterial cell wall and hundreds of new phage particles emerge. Each phage is now free to infect other host cells in the area and repeat the process.

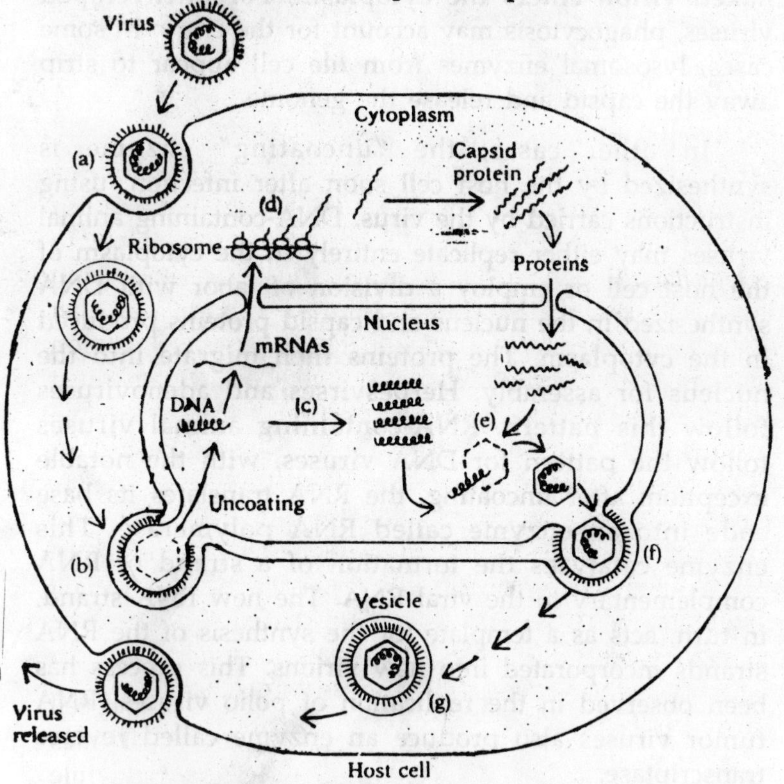

Fig. 2. Replication of DNA-containing animal virus inside

Method of replication displayed by animal viruses is similar to that of T-even phages with some notable exceptions. For example, the cell's adsorption site exists on the cell membrane rather than the cell wall Also, the adsorption sites on the virus are on the viral surface rather than on a tail. Influenza viruses, for instance, have sites on their spikes. As with phages, the adsorption and receptor sites are highly specific. This is why polio viruses attach to cells of the central nervous system and intestinal tract, but not other tissues. Penetration is also different. In animal viruses with envelopes, a fusion takes place between the envelope and cell membrane, and the naked virion enters the cytoplasm, For unenveloped viruses, phagocytosis may account for the entry. In some cases, lysosomal enzymes from tile cell appear to strip away the capsid and release the genome.

In other cases, the "uncoating" enzyme is synthesized by the host cell soon after infection, using instructions carried by the virus. DNA-containing animal viruses may either replicate entirely in the cytoplasm of the host cell or employ a division of labor with DNA synthesized in the nucleus and capsid proteins produced in the cytoplasm. The proteins then migrate into the nucleus for assembly. Herpesvirses and adenoviruses follow this pattern. RNA-containing animal viruses follow the pattern for DNA viruses, with the notable exception: after uncoating, the RNA translates its base code into an enzyme called RNA polymerase. This enzyme catalyzes the formation of a strand of RNA complementary to the viral RNA. The new RNA strand, in turn, acts as a template for the synthesis of the RNA strands incorporated into new virions. This process has been observed in the replication of polio viruses. RNA tumor viruses also produce an enzyme called reverse transcriptase.

Length of time spent in replication varies widely among animal viruses. For example, polio virses in nerve cells require 6 to 8 hours for a replication cycle, herpes simplex viruses in epidermal cells take 12 to 30 hours to replicate, and adenoviruses in respiratory cells spend 48 hours in a replication cycle, Enveloped viruses acquire their envelopes by budding through the cell membrane and generally do not destroy the host cells, but the most common response of the cell to viral replication is lysis.

Lysogeny

Infections of host cells by virious do not result in lysis. In these cases the viral genome codes for a repressor protein that prevents activity of the viral genes that would otherwise lead to replication. The nonmultiplying virus is called a temperate virus and the genome is known as a prophage. The prophage may attach to the host cell chromosome and multiply along with it for many generations resulting in phage conversion, or it may remain in the cytoplasm as a plasmid. The phenomenon is termed lysogeny, and the host cell is said to be lysogeny.

Classification

Like all other microorganisms, viruses have a set of characteristics that set them apart and help virologists understand their activities and deal with them effectively. In this section, we shall examine some of these characteristics.

A widely accepted classification system for viruses has not yet been devised, in part because insufficient data is available on viruses to determine their relationships to one another. Viruses therefore lack formal names. Instead, viruses have acquired their names from several sources. Examples are the measles virus (after the disease), the adenovirus (after the adenoids, where it is commonly located), the Coxsackievirus (after Coxsackie, New York, where it was originally isolated), and the Epstein-Barr virus (after researchers who studied it). The situation is reminiscent of a century ago when different bacteria were called the "tubercle bacillus" the "diphtheria bacillus"; and the cholera bacillus." Two schemes have been suggested for the classification of animal viruses. Third International Conference of Virology in Madrid. It recognizes 6 families of DNA viruses, 11 families of RNA viruses, and 1 family of miscellaneous viruses. The second classification scheme is loosely based on the body tissues that the viruses affect. Though inexact, this scheme places viruses into four convenient groups depending on whether they multiply in respiratory, skin, visceral, or nervous tissue.

Detection of viruses

Methods used to detect viruses are more involved and considerably more time-consuming than for bacteria and other microorganisms. Because viruses require host cells for replication, virologists must Provide some type of living tissue. Plant and animal tissues are difficult and expensive to maintain, and pathogenic viruses often replicate only in human host cells, a factor that causes additional complications. By contrast, bacteriophages are easily cultivated in bacterial cultures. This is one reason why bacteriophages have been used as models to study viral characteristics. One common method of detecting viruses is to inoculate them into fertilized eggs, A hole is drilled in the shell of the egg, and a suspension Of viral material is introduced. Because different viruses replicate in different membranes or parts of the chick embryo, virologists must anticipate which virus is present. Viral replication is detected by the death of the embryo cell damage to the embryo or its membrane, or formation of characteristic lesions at the site of inoculation.

To detect viruses tissue cultures method is also useful. To prepare the culture, cells are separated from a tissue with enzymes and suspended in a solution of nutrients, growth factors, pH buffers, and salts. The cells adhere to the wall of the container and reproduce to form a single layer, or monolayer. When viruses replicate in these cells, a noticeable deterioration occurs. This is called a *cytopathic effect (CPE).* Tissue culture is a highly sophisticated technique performed only in specialized laboratories.

Another method for detecting viruses is to search for viral antibodies in the serum of a patient. This can be done by combining a serum sample with known viruses. In some serological tests, the reaction is facilitated and made visible when the viruses are attached to a carrier particle. If the particles clump together, then the presence of viral antibodies can be inferred. Absence of clumping means that antibodies for that virus were absent.

Viruses such as influenza, measles, and mumps agglutinate (clump) red blood cells. This phenomenon, called *hemagglutination,* can be used to detect specific viruses. In addition, it is possible to detect antibodies against certain viruses because the antibodies will react with viruses and inhibit hemagglutination. Thus a laboratory test for antibodies can be performed by combining the patient's serum with known viruses and red blood cells. Hemagglutination indicates that antibodies are absent on the serum, but *emagglutination-inhibition* points to the presence of antibodies in the serum. This implies that the patient has been exposed to the viruses.

Viruses leave signs of their presence in infected tissue, in certain instances. For example, the brain cells of a rabid animal contain cytoplasmic granules called *Negri bodies,* and cells from herpes simplex patients have nuclear granules known as *Lipshutz bodies.* Moreover, a series of cellular or tissue modifications may signal the presence of viruses. Infectious mononucleosis, for example, is characterized by large numbers of swollen lymphocytes with foamy, highly vacuolated cytoplasm. Physicians call these cells *Downey cells.* Measles is accompanied by *Koplik spots,* a series of bright red patches with white pimplelike centers on the lateral mouth surfaces. Swollen salivary glands and teardroplike skin lesions arc associated with mumps and chickenpox, respectively.

Most obvious method for detecting viruses is by direct observation with the electron microscope. In this procedure, tissue samples may be examined directly or it may be necessary to increase the number of viruses by cultivation before samples are prepared for microscopy,

Virologists are often able to identify unknown viruses by comparison to known viruses. Bacterial viruses, or bacteriophages, may be detected by their formation of plaques. A plaque is a clear area on a "lawn" of bacteria where bacteriophages have destroyed the cells. Virologists first cultivate the bacteria on an agar surface, and then add the viruses by spraying.

If the viruses are specific for that bacterium, they infect and replicate in the cells and destroy them, thereby forming clear plaques within the cloudy bacterial surface. This method can also be reversed so that a known phage is used to detect an unknown bacterium. The diagnosis of several bacterial diseases such as staphylococcal food poisoning, is aided by this procedure, often called phage typing.

Difficulty in detecting viruses has a bearing on relating a particular virus to a particular disease. In the classical sense, Koch's postulates cannot be applied to a viral disease because the viruses do not grow in pure culture.

To circumvent this problem, Thomas M. Rivers in 1937 expanded Koch's postulates to include viruses by modifying the postulates as follows: Filtrates of the infectious material shown not to contain bacteria or other cultivatable organisms must produce the disease or its counterpart; or the filtrates must produce specific antibodies in appropriate animals. This concept is known as Rivers' postulate.

Inhibition of viruses

The human body attempts to rid itself of viruses mainly by phagocytosis and neutralization with antibodies, during periods of disease. Antibodies are protein molecules from the immune system that react specifically, with the virus that stimulated their formation. Usually one antibody molecule neutralizes a single virion thereby preventing the virion from penetrating a host cell. Antibodies also clump viruses into large masses that are efficiently removed by phagocytosis.

In addition, viruses activate the complement system, a group of substances that encourage phagocytosis. Normal body defenses to viral disease cannot be supplemented by antibiotics because viruses lack the structures and metabolism with which antibiotics interfere. For example, penicillin is useless for treating viral diseases because viruses have no cell wall.

However, there are several drugs that have been found effective against viruses. One drug, called *amantadine* (Symmetrel), was licensed by the Food and Drug Administration (FDA) for use against influenza viruses. Virologists believe it prevents attachment of viruses to the host cell surface. Another drug, *vidarabine* (Vira-A), has been in use for treating herpes zoster (shingles) and encephalitis (brain disease) due to the herpes simplex virus. Virologists have developed a valuable approach to viral inhibitlon in *base analogs*. These substances resemble nitrogenous bases and are erroneously incorporated into viral DNA. One example, *idoxuridine* (IDU), is taken up by herpesviruses in place of thymine, and the resulting genome cannot replicate itself. IDU has been available in ointment form for the treatment of herpes infections of the eye.

A new drug, *trifluridine*, works in a similar way but can be used in liquid form as eyedrops rather than an ointment; the eyedrop is preferable because the ointment blurs vision. *Interferon* represents one of the most optimistic approaches to the inhibition of viruses. First identified by Alick Isaacs and Jean Lindenmann, interferon is not a single substance but a group of over 20 substances designated alpha, beta, and gamma interferons. Each group has several members, and all appear to be proteins. Interferons are produced by various body cells on stimulation by viruses. They trigger a reaction that protects against the stimulating virus, as well as many other viruses. In addition, some interferons have anticancer properties. However, human interferon is the only one that will work in humans.

Interferons do not interact directly with viruses, but instead, with the cell they protect. For this reason they have a broader inhibitory effect. Interferons are produced when a virion releases its genome into the cell. The foreign material induces the cell to synthesize and secrete interferons. These bind to specific receptor sites on the surfaces of adjacent cells and trigger the production of several proteins within those cells.

The proteins inhibit viral replication by methods, not completely understood, although many virologists believe that at least one protein binds to messenger RNA molecules coded by the virus. The inability to obtain sufficient amounts for research stifled interferon research for many years. One approach was to inoculate huge batches of white blood cells with harmless viruses. However, a breakthrough when Swiss and Japanese scientists deciphered the genetic code for interferon, and spliced E. coli plasmids *Plasmi* with the DNA code.

Experiments showed that interferon from bacterial factories would reduce hepatitis symptoms, diminish the spread of herpes zoster, and shrink certain cancers. Swiss biotechnology firm received a patent for alpha interferon and began marketing its product using the trade name Intron.

Viral vaccines

Vaccines stimulate antibody production by the body's immune system and thereby induce specific defense to disease. Two types of viral vaccines are currently in use.

The first contains *inactivated viruses* treated with a physical agent such as heat, or a chemical agent such as formaldehyde. This treatment alters the genome of the virus, thereby preventing replication, but it does not severely affect the capsid, which remains intact and stimulates antibody production. Opponents of inactivated viral vaccines point out that treatment may not reach all viruses and that untreated viruses may remain active and cause disease.

Another type of vaccine contains *attenuated viruses* that continue to replicate in body cells but at an extremely low rate. Attenuated viruses stimulate the immune system for a longer period of time than inactivated viruses. They are obtained by transferring viruses from culture to culture for a period of months or years until a variant emerges with a greatly reduced replication rate. Opponents of attenuated viral vaccines suggest that the viruses may revert back to their original form and cause disease, and that the viruses may infect cells other than the usual host cells or activate temperate viruses already in host cells.

Viruses and cancer

Cancer is a condition resulting from the uncontrolled reproduction of cells. The cells in a cancer do not multiply at a faster rate than normal cells but they do multiply more frequently as the time between divisions diminishes. Thev escape controlling factors and eventually form a cluster of cells. Soon the cluster becomes an abnormal functionless mass of cells called a *tumor*. The body normally responds to a tumor by surrounding it with a capsule of connective tissue. Such a tumor is said to be *benign*. However, if the cells multiply too rapidly to be restricted, or if they break out of the capsule and spread (metastasize), the tumor is described as *malignant*. The condition is now called cancer, meaning that the cells are radiating out like the arms of a crab.

The study of cancer is called *oncology*, from the Greek *oncus* for tumor. Cancer cells differ from normal cells in three major ways: they multiply more frequently than normal cells; they fail to adhere tightly to one another (as normal cells do) and overgrow one another; and they revert to the formless cells of their youth and divide rapidly like embryonic cells. There appears to be no boundary limiting a cancer. The cells infiltrate a broad variety of tissues, and since they fail to stick to tissues, they spread almost anywhere. A *fibroma*, for example, is a tumor of the muscle tissue; a *papilloma* is a tumor of the skin tissue, often called a wart; a *sarcoma* is a tumor of the connective tissue; and *leukemia* is a cancer of the white blood cells.

Cancer cells interfere with an organ's functions by the sheer force of their numbers. They rob normal cells of nutrients and induce cell death by starvation. Also, they clog the vascular network and cut off the oxygen supply to cells. The immune system is apparently unable to control cell proliferation in tumors and cancers. Cancer-causing substances are called *carcinogens*. The list of carcinogens includes chemicals such as hydrocarbons in cigarette tar, and physical agents such as ultraviolet light in sunlight.

Considerable controversy exists on whether viruses are carcinogens, with most evidence indicating an involvement in cancer. Francis Peyton Rous postulated that a virus was involved in sarcomas in chickens. Edwin Shope suggested that a virus was causing papillomas in rabbits; and Denis Burkitt reported that lymphomas (tumors of the lymph tissue) were caused by a virus. Experiments with animals indicate that viruses do, indeed, induce tumor formation. However, federal law and ethical standards prohibit these experiments from being repeated in human volunteers because an effective treatment for cancer is not yet available.

Nevertheless, a number of viruses have been isolated from human cancers and, when these viruses are transferred to tissue cultures and animals, an observable cell transformation takes place. Examples of human tumor viruses include types of adenoviruses, echoviruses, and herpesviruses. Animal tumor viruses include the Simian virus 40 (SV40), which affects kidney tissues in monkeys, the Shope papilloma virus of rabbits, the leukemia viruses of cats, dogs, and hamsters, and the rnouse mammary tumor virus, discovered by John Joseph Bittner. Bittner's work was notable because it showed that the virus could be transmitted among animals. How viruses trigger cells to become tumor cells has been the subject of intensive research in recent years.

Most studies have centered about tile *oncogene theory*, first proposed by Robert Huebner and George Todaro. This theory suggests that the genes to transform a cell, the *oncogenes*, reside in the chromosomal DNA of a normal cell. When a genetic change occurs, such as by introduction of a virus, the oncogenes transform the cell to a cancer cell. In the modern view of cancer researchers, the oncogenes are presumed to exist as genes called *protooncogenes.* Proto-oncogenes may have important functions in the normal regulation of cell growth and development, perhaps by turning on at specific stages of life and turning off later.

Indeed, research reported in linked the proto-oncogenes to the production of cyclic adenosine monophosphate (cAMP), an organic substance central to the regulation of numerous physiological processes. A clue to the importance of proto-oncogenes comes from their observation in diverse forms of life from yeasts to humans. This finding indicates that proto-oncogenes have not changed during the millions of years of evolution. Oncologists believe that chromosomal alterations by cancer agents convert proto-oncogenes into oncogenes that transform a cell.

Virologists have observed that when a virus enters the relationship of lysogeny with a cell, it may transform the cell. If the viral genome is composed of DNA, the DNA attaches directly to the chromosome like a prophage attaches to the chromosome of a bacterial cell. However, if the genome is composed of RNA, then an enzyme called *reverse transcriptase* synthesizes DNA from RNA. Once the synthesis is completed, the new DNA binds to the cell chromosome. One mechanism of transformation has been studied in a group of RNA viruses called *retroviruses.* These viruses are so-named because their reverse transcriptase "reverses" the flow of genetic information from RNA to DNA.

The retroviruses carry oncogenes, which are inserted onto a host cell's chromosome along with the newly synthesized viral DNA. This process is akin to transduction. Once on the host cell chromosome, the oncogenes apparently "switch on" the cell's proto-oncogenes and activate them to oncogenes that transform the cell. Robert Gallo and his co-workers at the National Cancer Institute identified a retrovirus as the cause of a cancer of the white blood cells called *human T-cell leukemia.* The virus, believed to be the first linked to human cancer, was named HTLV-1. Gallo and his group identified another retrovirus, HTLV-III, as the agent of AIDS.

Another method of transformation has been studied with the virus of *Burkitt's lymphoma.* In this cancer of the lymphoid connective tissues of the jaw, the viral genome appears to insert itself into a chromosome of B-lymphocytes, the white blood cells important in immunity. The insertion triggers certain proto-oncogenes involved in cell growth to move from their position on chromosome 8 to a new position on chromosome 14, far from the influence of their control genes. A segment of DNA from chromosome 14 replaces the proto-oncogenes from chromosome 8. The proto-oncogenes, now oncogenes, appear to produce elevated amounts of their protein product, which may be related to abnormally rapid cell growth and reproduction. An important practical benefit of this finding is that the chromosome change can be detected in the cancer patients, thus aiding diagnosis.

Over 20 oncogenes have been discovered in human cancers including those of the bladder, lung, and colon, and a number of the oncogenes have been linked to the production of various proteins. One series of studies conducted at Cold Spring Harbor Laboratories in New York was particularly interesting.

A team led by Michael Wigler isolated a protein whose production was directed by an oncogene. The protein was injected into normal cells and within hours, the normal cells began to multiply rapidly and show unmistakable signs of conversion to cancer cells. Growth slowed down several hours later, presumably because the supply of protein was exhausted. In a later experiment, they injected antibodies against the protein into cancer cells and observed signs of reversion to normal cells. The experiments strengthened the evidence that abnormal products of oncogenes enhance cell transformation and that, in theory, countermeasures are possible. Although researchers are quick to point out that cancer cures are not in the near future, they note that oncogene studies may provide an understanding of how to prevent cancers.

Knowledge of oncogenes may also aid the early detection of cancers and assist the development of anticancer drugs. The oncogene theory sets down a unifying mechanism by which all carcinogens may act regardless of whether they are chemical substances, physical agents, or viruses.

Viroids

Viroids are tiny fragments of nucleic acid known to cause several diseases in plants and thought to be involved in human and animal diseases. The discovery of viroids resulted from studies conducted at the U.S. Department of Agriculture's Beltsville, Maryland, research center near Washington, D.C. Scientists led by Theodore O. Diener were investigating a suspected viral disease, *potato spindle tuber (PST)*, which results in small, cracked potatoes shaped like spindles. Nothing would destroy the disease agent except an RNAdissolving enzyme, the group postulated that a fragment of RNA was involved. Diener called the agent a viroid, meaning viruslike. The next year, a team led by Joseph Semancik at the University of California found a similar agent in a disease of citrus trees.

There are at least a dozen plant diseases have been related to viroids. The largest of these particles is about one-twentieth of the size of the smallest virus. The RNA chain of the PST viroid has a known molecular sequence, but it contains so few genetic codes that the replication cycle is not understood. Diener has speculated that the viroids originated as introns, the sections of RNA spliced out of messenger RNA molecules before the messengers are able to function. The similarity in size between introns and viroids and the ring shape for both has fueled the speculation.

Semancik has offered the theory that viroids may be regulatory genes because viroid diseases are characterized by interference with plant growth. An interesting twist in viroid research occurred when it was found that viroids could infect animals. Researchers from Knoxville, Tennessee, reported DNA viroids in hamster colonies afflicted with lymphatic cancer. Currently, some microbiologists believe that viroids may cause the mysterious slow virus diseases of humans.

Chapter 4

Control of Microorganisms

The idea to control disease caused by microorganisms had been proposed by Pasteur decades before, but not until the turn of the century did it gather momentum and achieve a firm footing in the scientific community. Two types of control gradually evolved: physical control and chemical control. To achieve physical control, scientists employed heat, radiations, filters, and other physical agents to remove microorganisms from instruments, equipment, and fluids. Pasteurization of dairy products was introduced and food preservation methods were gradually updated. To achieve chemical control, doctors utilized antiseptics and disinfectants in medicine, surgery, and the treatment of wounds. Control of microorganisms is an essential factor to maintaining good health.

Physical control

sterilization methods

Sterilization implies the destruction or removal of all life forms. It is an absolute term that cannot be qualified. Thus one cannot assume "partial sterilization" or incomplete sterilization," but must consider a material "contaminated" until sterilized. It should be pointed out, however, that material may remain harmful even though it is sterile. For example, a solution of bacterial toxin may contain no living forms but still cause damage in the body.

Although chemicals may sometimes be used to sterilize objects, the principal methods for achieving sterilization employ physical agents such as heat, as well as radiations, and filtration. These methods art-not products of the modern era. They were used by Pasteur, Koch, and microbiologists of a century ago to prevent contamination of their materials and to ensure the accuracy of their work. In this section we shall discuss many of these physical methods, at the same time noting how they are used to interrupt the chains of microbial transmission.

Bacterial spores are the most resistant forms of life. Anthrax spores, for example, have been found viable after 60 years on dry silk threads. The spores are not metabolically inert. Rather, they carry on life processes at minimum rates and possess the necessary enzymes to transform from a dormant state to actively metabolizing vegetative cells.

The heat resistance of spores is often attributed to their low water content, which yields a heat-resistant get like spore core. Another possible reason is the presence of dipicolinic acid, an organic compound that encourages heat resistance by linking to the spore proteins.

Destruction of the bacterial spore is the principal aim of sterilization methods, especially those involving heat. Microorganisms grow over a wide variety of temperatures, but growth slows down above and below the optimal temperature. Above maximum growth temperatures, biochemical changes in the organic molecules result in cell death. These changes arise from alterations in enzyme molecules or chemical breakdowns of structural molecules, especially in the cell membranes.

Heat also drives off water, and since all organisms depend on water, this loss may be lethal. The killing rate of heat may be expressed as a function of time and temperature. For example, tubercle bacilli are destroyed in 30 minutes at 58° C, but only 2 minutes at 65°C, and a few seconds at 72°C. Each microorganism has a thermal death time, the time necessary for killing it at a given temperature.

Each microorganism also has a thermal death point, the temperature at which it dies in a given time. These measurements are particularly important in the food industry, where heat is used for preservation. In determining the time and temperature for microbial destruction with heat, certain factors bear consideration.

One factor is the type of organism to be killed. For example, if materials are to be sterilized, the physical method must be directed at bacterial spores. Milk, however, need not be sterile for consumption, and heat is therefore aimed at the most resistant vegetative cells. Another factor is the type of material to be treated. Powder, for instance, is subjected to dry heat because moist heat will leave it soggy.

Saline solutions, by contrast, can be sterilized with moist heat but are not easily treated with dry heat. Other factors are the presence of organic matter and the acidic or basic nature in the material. Organic matter may prevent heat from reaching microorganisms, while acidity or alkalinity-may assist the lethal action of heat.

Perhaps the most rapid sterilization method is the direct flame method used in the process of incineration, A flame is employed to sterilize the bacteriological loop in the Bunsen burner before removing a sample from a culture tube and after preparing a smear. Flaming the tip of the tube also destroys organisms that happen to contact the tip, white burning away lint and dust in general, objects must be disposable if a flame is used for sterilization. Disposable hospital gowns and certain plastic apparatus are examples of materials that may be incinerated. In past centuries, the bodies of disease victims were burned to prevent spread of the pestilence. It is still common practice to incinerate the carcasses of cattle that have died of anthrax and to put the contaminated field to the torch because anthrax spores cannot adequately be destroyed by other means. British law stipulates that anthrax- contaminated animals may not be autopsied and must be burned.

Hot-air oven utilizes radiating dry heat for sterilization. This type of energy does not penetrate materials easily and thus, long periods of exposure to high temperatures are necessary. For example, at a temperature of 160°C (320° F), a period of two hours is required for the destruction of bacterial spores. Higher temperatures are not recommended because the wrapping paper used for equipment tends to char. The hot-air method is useful for the sterilization of dry powders and water-free oily substances, as well as many types of glassware such as pipettes, flasks, and syringes.

Dry heat does not corrode sharp instruments as steam often does, nor does it erode the ground glass surfaces of syringes. The effect of dry heat on microorganisms is equivalent to that of' baking. The heat changes microbial proteins by oxidation reactions and creates an arid internal environment, thereby burning microorganisms slowly. It is essential that organic matter such as oil or grease films be removed from the materials, because organic matter insulates against dry heat. Moreover, the time required for heat to reach sterilizing temperatures varies among materials.

Immersion in boiling water is the first of several moist-heat methods that we shall consider. Moist heat penetrates materials much more rapidly than dry heat because water conducts heat better than air. Lower temperatures and less time of exposure are therefore required than for dry heat. Moist heat kills microorganisms by denaturing their proteins Denaturation involves changes in the chemical or physical properties of proteins and includes structural alterations due to destruction of the chemical bonds that hold proteins in a three-dimensional form. A," proteins revert to a two-dimensional structure, they coagulate (precipitate) and become nonfunctional. Egg protein undergoes a similar transformation when it is boiled. You might find a review of the chemical structure of proteins in helpful to your understanding of this process. The coagulation and denaturing of proteins require less energy than oxidation, and therefore, less heat can be used.

Under ordinary circumstances, with microorganisms at concentrations of less than one million per millilitre, most microorganisms can be killed within 10 minutes. Indeed, a few seconds may be all that is necessary. However, fungal spores, protozoan cysts, and large concentrations of hepatitis A viruses require up to 30 minutes' exposure, Bacterial spores often require two hours or more. Because inadequate information exists on the heat tolerance of many microorganisms, boiling water is not reliable for sterilization purposes.

If it is imperative that boiling water be used to destroy microorganisms, materials must be thoroughly cleaned to remove traces of organic matter such as blood or faces. The minimum exposure period should be 30 minutes, except at high altitudes, where it should be increased to compensate for the lower boiling point of water. All materials should be well covered. Washing soda may be added at a 2 percent concentration to increase the efficiency of the process.

Moist heat in the form of pressurized steam is regarded as the most dependable method for the destruction of all forms of life, including bacterial spores. This method is incorporated into a device called the autoclave. Over a hundred years ago, French and German microbiologists developed the autoclave as an essential component of their laboratories. A basic tenet of chemistry is that when the pressure of a gas increases, the temperature of the gas increases proportionally. Because steam is a gas, increasing its pressure in a closed system increases its temperature. As the water molecules in steam become more energized, their penetration increases substantially. This principle is used to reduce cooking time in the home pressure cooker and to reduce sterilizing time in the autoclave.

The sterilizing agent is the moist heat, not the pressure. Most autoclaves contain a sterilizing chamber into which articles are placed and a steam jacket where steam is maintained, as steam flows from the steam jacket into the sterilizing chamber, cool air is forced out and a special valve increases the pressure to 15 pounds/square inch (lb/in^2) above normal atmospheric pressure The temperature rises to 121.5° C and the superheated water molecules rapidly conduct heat into microorganisms. The time for destruction of the most resistant bacterial spore is now reduced to about 15 minutes. For objects that are denser, up to 30 minutes of exposure may be required.

The autoclave is used for the control of microorganisms in both hospitals and laboratories. It is employed for blankets, bedding, utensils, instruments, intravenous solutions, and a broad variety of other objects. The laboratory technician uses it to sterilize bacteriological media and destroy pathogenic cultures. It is equally valuable for glassware and metalware, and is among the first instruments ordered when a microbiology laboratory is established. The autoclave also has certain limitations. For example some plasticware melts in the high heat, and sharp instruments often become dull.

Moreover, many chemicals break down during the sterilization process, and oily substances cannot be treated since they do not mix with water. In addition, the various times recommended for different objects must be observed, and objects must be separated within the chamber to assure penetration of the superheated steam. To gauge the success of sterilization, a strip containing spores of a *Bacillus* species is included with the objects treated. At the conclusion of the cycle, the strip is placed in nutrient broth medium. If the sterilization process has been successful, no growth will occur. Growth of the Bacillus indicates that the sterilization process has failed.

A new form of autoclave called the prevacuum autoclave, has been developed for sterilization procedures. This machine draws air out of the sterilizing chamber at the beginning of the cycle. Saturated steam is then used at a temperature of 132°C to 134°C at a pressure of 28 to 30 lb/in^2. The time for sterilization is now reduced to as little as 4 minutes. A vacuum pump operates at the end of the cycle to remove the steam and dry the load.

The major advantages of the prevacuum, autoclave are the minimal exposure time for sterilization and the vacuum to reduce the time for completion of the cycle. In the years before development of the autoclave, liquids and other objects were sterilized by exposure to free-flowing steam at 100°C for 30 minutes on each of three successive days. The method was called fractional sterilization because a fraction was accomplished on each day. It was also called tyndallization after its developer, John Tyndall and intermittent sterilization because it was a stop-and start operation.

Fractional sterilization method is achieved by an interesting series of events. During the first day's exposure, steam kills virtually all organisms except bacterial spores. When the material is incubated overnight, the spores germinate into vegetative cells, which are killed during the second day's exposure to steam. Again, the material is cooled and the few remaining spores germinate, only to be killed on the third day.

Although the method usually results in sterilization, occasions arise when several spores fail to germinate, and therefore, sterility of the material cannot be assured. The method also requires that spores be in a suitable medium for germination, such as a broth. Fractional sterilization has assumed importance in modern microbiology with the development of high-technology instrumentation and new chemical substances. Often, these materials cannot be sterilized at autoclave temperatures, or by long periods of boiling or baking, or with chemicals. An instrument that generates free-flowing steam, such as the Arnold sterilizer, is used in these instances.

Pasteurization

Pasteurization aims to reduce the bacterial population of a liquid such as milk and destroy organism that may cause spoilage and human disease. Spores are not affected by pasteurization. One method for milk pasteurization, called thc holding method, involves heating at 62.9°C for 30 minutes. Although thermophilic bacteria thrive at this temperature, they are of little consequence because they cannot grow at body temperature.

For decades, pasteurization has been aimed at destroying *Mycobacterium tuberculosis, long* considered the most heat-resistant bacterium. More recently, however, attention has shifted to destruction of *Coxiella burnetii*, the agent of Q fever, because this organism has a higher resistance to heat. Since both organisms are eliminated by pasteurization, dairy microbiologists assume that other pathogenic bacteria are also destroyed.

Two other methods of pasteurization are the flash pasteurization method at 71.6° C for 15 seconds, and the ultrapasteurization method at 82° C for 3 seconds. Some dentists and physicians use hot oil at 160°C for the sterilization of instruments. A time period of 1 hour is usually recommended. This method is advantageous because oil does not rust metals, and thus, minimal corrosion takes place. However, after sterilization, the instruments must be cleaned and dried thoroughly for storage and this step may reintroduce contamination.

Filtration

During early peroids had been utilized to trap airborne organisms and sterilize bacteriological media, but now they became essential for separating viruses from other microorganisms. Among the early pioneers of filter technology was Charles Chamberland, an associate of Pasteur. Another pioneer was Julius Petri (inventor of the Petri dish), who developed a sand filter to separate bacteria from the air. The filter is a mechanical device for removing microorganisms from a solution. As fluid passes through the filter, organisms are trapped in the pores of the filtering material. The solution that drips into the receiving container is decontaminated or, in some cases, sterilized. Such things as intravenous solutions, certain bacteriological media, toxoid preparations, and numerous other liquids may be treated in this way.

There are several types of filters available for use in the microbiology, laboratory. Inorganic filters are typified by the Seitz filter, which consists of a pad of porcelain or ground glass mounted in a filter flask. Organic filters are advantageous because the organic molecules of the filter attract organic components in microorganisms. One example, the Berkefeld-filter, utilizes a substance called diatomaceous earth. This material is composed of the remains of marine algae known as diatoms. Diatoms are unicellular algae that abound in the oceans and provide important foundations for the world's food chains. Their remain accumulate on the shoreline and are gathered for use in swimming pool and aquarium filters, as well as for microbiological filters used ill laboratories.

Membrane filter is a third type of filter that has received broad acceptance. It consists of a pad of organic compounds such as ccllulose acetate or, polycarbonate, mounted in a holding device. This filter is particularly valuable because bacteria multiply and form colonies oil the filter pad when the pad is placed on a plate of culture medium.

Microbiologists can then count the colonies to determine the number of bacteria originally present. For example, if a 100 ml sample of liquid were filtered and 59 colonies appeared on the pad after incubation, it could be assumed that 59 bacteria were in the sample. Membrane filters are available in various pore sizes selected according to which microorganisms are to be trapped. Different type of organism may be counted by varying the culture medium, and the filed micro-biologists can adapt the filter system to a portable laboratory. The membrane filter also has various industrial uses for sterilizing fluids and certain beverages as points out. Special sieve-type membrane filters may be used to trap microorganisms for identification by electron microscopy.

Ultraviolet light

Ultraviolet light has a wavelength of between 100 and 400 nm, with the energy at about 265 nm most destructive to bacteria. When microorganisms are subjected to ultraviolet light, cellular DNA absorbs the energy, and adjacent thiamine molecules link together. Thymine molecules thus become unable to position adenine on messenger RNA molecules during the process of protein synthesis. Also replication of the chromosome in binar fission is impaired. The mutated organism can no longer produce certain proteins or reproduce, and it dics quickly. Ultraviolet light is valuable for reducing the microbial population where direct exposure takes place. It is used to reduce airborne or surface contamination in a hospital room, morgue, pharmacy, toilet facility, or food service operation. It is noteworthy that ultraviolet light from the sun may be an important factor in controlling microorganisms in the air and upper layers of the soil. One drawback to practical use is its failure to penetrate liquids or solids. Another is the damage ultraviolet light may cause in human skin cells.

The energy spectrum includes two other forms of radiation useful for destroying microorganisms. These are X rays and gamma rays. Both have wavelengths shorter than the wavelength of ultraviolet light. As X rays and gamma rays pass through microorganisms, they force electrons out of their shells in organic molecules, thereby creating ions. For this reason, the radiations are called ionizing radiations. The ions attract one another and form unusual molecules that change the structure or chemistry of the microorganism. Occasionally, toxic chemicals such as peroxides may result from the reaction. Ionizing radiations are currently used to sterilize such heat-sensitive pharmaceuticals as vitamins, hormones, and antibiotics as well as certain plastics and suture materials.

The microwave, has a wavelength longer than that of ultraviolet light. In a microwave oven, microwaves are absorbed by water molecules, The molecules are set into high-speed motion, and the heat of friction is transferred to foods, which become hot rapidly. Other than the heat generated, there is no specific activity against microorganisms. A final form of radiation that we shall consider is light energy.

When concentrated by sophisticated devices, light energy forms a laser beam. The term "laser" stands for light amplification by stimulated emission of radiation. Recent experiments have indicated that laser beams can be used to sterilize instruments and the air in operating rooms, as well as to kill organisms on a wound surface. Microorganisms are destroyed in a fraction of a second, but the laser beam must reach all parts of the material to effect sterilization.

Ultrasonic vibrations

Ultrasonie-vibrations are high-frequency sound waves. When directed against environmental surfaces, they have little value because air particles deflect and disperse the vibrations. However, when propagated in fluids, ultrasonic vibrations cause the formation of microscopic bubbles, or cavities, and the water appears to boil. Some observers call this "cold boiling." The cavities rapidly collapse, and send out shock waves. Microorganisms in the fluid are quickly disintegrated by the external pressures. Ultrasonic vibrations are valuable in research for breaking open tissue cells and obtaining their parts for study. A device called the cavities is used by dentists to clean teeth, and ultrasonic machines are available for cleaning dental plates, jewelry, and coins. A major appliance company has also experimented with an ultrasonic washing machine. As a sterilizing agent, ultrasonic vibrations have received minimal acceptance because liquid is required and other methods are more efficient. However, many research laboratories use ultrasonic probes for cell disruption and hospitals use ultrasonic devices to clean their instruments. When used with an effective germicide, an ultrasonic device may achieve sterilization, but the current trend is to use ultrasonic vibrations as a cleaning agent and precede the process by sterilization in an autoclave.

Preservation methods

Various physical methods have evolved for controlling microorganisms in food. Though valuable for preventing the spread of infectious agents, these procedures are used principally to retard spoilage and prolong the shelf life of foods rather than for sterilization.

Drying is useful in the preservation of various meats, fish, cereals, and other foods, Since water is a necessary requisite for life, it follows that where there is no water, there is virtually no life. Many of the foods in the kitchen pantry typify this principle.

Salting is based upon the principle of osmotic pressure. When food is salted, water diffuses out of microorganisms to dilute the salt in the surrounding environment. This flow of water, called osmosis, causes microorganisms to shrivel and die. The same phenomenon occurs in highly sugared foods. This accounts for the preservation of syrups, jams, and jellies from bacterial contamination.

At loow temperatures refrigerator and freezer help to prevent spoilage by reducing the rate of metabolism in microorganisms and, consequently, reducing their rate of growth. Spoilage is not totally eliminated in cold foods, however, and many microorganisms remain alive, even at freezer temperatures. These organisms multiply rapidly when food thaws, which is why prompt cooking is recommended.

Chemical control

Over the centuries, necessity demanded chemicals for food preservation, and spices were effective as preservatives as well as masks for foul odors. Marco

Polo's trips to the Orient for new spices were thus borne out of necessity as well as adventure. Medicinal chemicals came into increased. For example, the U.S. *Pharmacopeia* listed tincture of iodine as a valuable antiseptic, and soldiers in the Civil War used it widely. In the first decades of the century, people found copper sulphate useful for preventing fungal disease in plants, but unfortunately; this chemical was not so widely accepted to be available during the great potato blight of Ireland. Mercury was sometimes used for treating syphilis, as first suggested by Arabian physicians centuries before. Moviegoers have probably noted that American cowboys practiced disinfection by pouring whiskey onto gunshot wounds between drinks.

Pasteur's references to airborne microorganisms convinced that microorganisms were the cause of wound infections. Lister experimented with several chemicals to kill microorganisms and finally settled on carbolic acid. He applied the chemical to wounds and instruments and sprayed it in the air near the operating table. Lister achieved brilliant success and was accorded many accolades as a pioneer microbiologist of his time, although be advocated the same principles recommended by Semmelweis years before.

Chemical control of microorganisms extends into such diverse areas as the hospital environment, food processing plant, and everyday household. These broad fields have yielded an equally broad terminology that should be explored before we undertake a discussion of individual chemical agents. The physical agents for controlling microorganisms are generally intended to achieve sterilization. This implies the destruction of all forms of life, especially bacterial spores. Chemical agents, by contrast, rarely achieve sterilization. Instead, they are expected only to destroy the pathogenic organisms on or in an object.

The process of destroying pathogens is called disinfection; the object is said to be disinfected. If the object is lifeless, the chemical agent is known as a disinfectant. However, if the object is living, such as a tissue of the human body, the chemical is an antiseptic. Antiseptics and disinfectants are usually bactericidal, but occasionally they may be bacteriostatic. A bactericidal agent kills microorganisms, while a bacteriostatic agent temporarily prevents their further multiplication without necessarily killing them. For example, a bactericidal chemical may inactivate the enzymes of an organism and interfere with its metabolism so that it dies,

A bacteriostatic chemical, by contrast, disrupts a minor chemical reaction and slows the metabolism ' resulting in a longer time between divisions. Although a delicate difference sometimes exists between the bactericidal and bacteriostatic nature of a chemical, the terms indicate effectiveness of the chemical agent in a particular situation.The word sepsis is derived from the Greek seps, meaning putrid. It refers to contamination of an object by microorganisms and forms the stems for "septicemia," meaning microbial infection of the blood, and "antiseptic," which translates to "against infection."

It is also the origin of the term *aseptic*, meaning free of contaminating microorganisms. Other expressions associated with chemical control are *sanitize* and to reduce the microbial population to a safe level as determined by local public health standards. For example, in diary and food processing plants, the equipment is usually sanitized. To *degerm* something is merely to remove organisms from a surface. For instance, washing with soap and water degerms the skin surface but has little effect on microorganisms deep in the skin pores. Other group of terms that bears mention is the -cidal agents. These include the fungicidal agents, which kill fungi; the virucidal agents, for viruses; the sporicidal agents, for bacterial spores; and the germicidal agents, for various types of microorganisms.

Antiseptics selection

A chemical agent must have certain properties, some of which are more desirable than others, to be useful as an antiseptic or disinfectant. The first prerequisite is that it must be able to kill microorganisms. It should also be nontoxic to animals or humans, especially if it is used as an antiseptic. It should be soluble in water and have a substantial shelf life during which its activity is retained. The agent should be useful in very diluted form and perform its job in a relatively short time. Both factors tend to reduce toxic side effects. Other characteristics will also contribute to the value of a chemical agent: it should not separate on standing; it should penetrate well; and it should not corrode instruments.

The chemical will have a distinct advantage. If it does not combine with organic matter such as blood or feces, because microorganisms are often found in this type of material. Of course, the chemical should be easy to obtain and relatively inexpensive. Since disinfection is essentially a chemical process, the parameters of chemistry should be considered when selecting an antiseptic or disinfectant. For example, the temperature at which the disinfection is to take place may be important because a chemical reaction occurring at 37° C (body temperature) may not occur at 25°C (room temperature). Also, a particular chemical may be effective at a certain pH but not another.

Moreover, the chemical reaction may be very rapid with one agent and slower with another. Thus if long-term disinfection is desired, the second agent may be preferable. Two other considerations are the type of microorganism to be eliminated and the surface treated. For instance, the removal of bacterial spores requires more vigorous treatment than the removal of vegetative cells. Also, a chemical applied to a laboratory bench is considerably different than one used on a wound. It is therefore imperative distinguish the antiseptic or disinfectant nature of a chemical before proceeding with use.

Table 1

Phenol coefficients of some common antiseptics and disinfectants

Chemical agent	*Staphylococcus aureus*	*Salmonella typhi*
Phenol	1.0	1.0
Chloramine	133.0	100.0
Tincture of iodine	6.3	5.8
Lysol	5.0	3.2
Mercury chloride	100.0	143.0
Ethyl alcohol	6.3	6.3
Formalin	0.3	0.7
Hydrogen *peroxide*	-	0.01

The United States Department of Agriculture (USDA) lists over 8000 disinfectants for hospital use and thousands more for general use. Evaluating these chemical agents is a tedious process because of the broad diversity of conditions under which they are used. A standard of effectiveness used for the chemical agents is the phenol coefficient (PC). This is a number that indicates the disinfecting ability of an antiseptic or disinfectant in comparison to phenol under identical conditions.

Phenol coefficient is determined by a laboratory procedure in which dilutions of phenol and the test chemical are mixed with standardized bacteria such as Staphylococcus aureus and *Salmonella typhi*. The laboratory technician then determines which dilutions have killed the organisms after a 10-minute exposure but not after a 5-minute exposure. The reciprocal of the dilution of the chemical is then divided by the reciprocal of the dilution of phenol to obtain the phenol coefficient for the chemical. The phenol coefficient test has many drawbacks, such as its determination in the laboratory rather than in a real-life situation. The test does not take into account such factors as tissue toxicity, activity in the presence of organic matter, or variations of temperature. Nevertheless, the phenol coefficient test remains one of the few laboratory tests available for evaluating chemical agents.

Another method is to dry standardized cultures of bacteria within small stainless steel cylinders and then expose the cylinders to the test chemical, After an established time, the bacteria are assayed for survival to the chemical. These methods of standardization have value under certain circumstances, but it is conceivable that a universal test may never be developed in view of the huge variety of chemical agents available and the numerous conditions under which they arc employed.

Chemical agents currently in use for controlling microorganisms range from the very simple substances such as halogen ions to the very complex compounds typified by the detergents. Many of these agents have been employed for generations, while others represent the latest developments. In this section we shall survey several groups of chemical agents and indicate how they are best applied in the chemical control of microorganisms.

Halogens

Halogens are a group of highly reactive elements whose atoms have seven electrons in the outer shell. Two halogens, chlorine and iodine, are commonly used for disinfection.

Chlorine

Chlorine is widely used in municipal water supplies, where it keeps bacterial populations at low levels. Chlorine combines readily with numerous ions in water and therefore, enough chlorine must be added to ensure that a residue remains for antibacterial activity. In municipal water, the residue of chlorine is usually about 0.2 to 1.0 parts per million (ppm) of free chlorine. One ppm is equivalent to 0.0001 percent, an extremely small amount. Chlorine is also available as *sodium hypochlorite* (NaOCl) or as *calcium hypochlorite* ($Ca(OCl)_2$). The latter, also known as chlorinated lime, was used by Semmelweis in his studies in Vienna. Sodium hypochlorite is used as a bleaching agent in the textile industry; commercially available bleach contains about 5 percent of this compound.

To disinfect clear water, the Centers for Disease Control recommends a half-teaspoon of household chlorine bleach in 2 gallons of water, with 30 minutes contact time before consumption. Hypochlorites are also useful in very dilute solutions for disinfecting swimming pools and sanitizing factory equipment. The *chloramines* such as chloramine-T are organic compounds that contain chlorine. These compounds release free chlorine more slowly than hypochlorite solutions and are more stable. They are valuable for general wound antisepsis and root canal therapy. Chlorine is effective against a broad variety of organisms including most Gram-positive and Gram-negative bacteria, and many viruses, fungi, and protozoa. However, it is not sporicidal.

In microorganisms, the halogen is believed to cause the release of atomic oxygen, which then combines with and inactivates certain cytoplasmic proteins such as enzymes. Another theory is that chlorine changes the structure of cell membranes, thus leading to leakage.

Iodine

Iodine atom is slightly larger than the chlorine atom and is more reactive and more germicidal. It is widely found in nature in such plants as marine seaweeds, where it is bound to chemical compounds. Iodine acts by halogenating tyrosine portions of protein molecules. *Tincture of iodine*, a commonly used antiseptic for wounds, consists of 2 percent iodine and sodium iodide dissolved in ethyl alcohol.

For the disinfection of clear water, the CDC recommends 5 drops of tincture of iodine in 1 quart of water, with 30 minutes contact time before consumption. Iodine compounds in different forms are also valuable sanitizers for restaurant eating utensils and equipment. Iodophors are iodine-detergent complexes that release iodine over a long period of time and have the added advantage of not staining tissues or fabrics. The detergent portion of the complex loosens the organisms from the surface and the halogen kills them.

Some examples of iodophors are Wescodyne, used in preoperative skin preparations; Ioprep for presurgical scrubbing; Iosan, for dairy sanitation; and Betadine, for local wounds. Iodophors may also be combined with nondetergent carrier molecules. The best known carrier is povidone, which stabilizes the iodine and releases it slowly. However, compounds like these are not self-sterilizing.

Phenolic compounds

Phenol and its compounds have played a key role in disinfection practices since Joseph Lister used them. Phenol remains the standard against which other antiseptics and disinfectants are evaluated in the phenol coefficient test. It is active against Grampositive bacteria, but its activity is reduced in the presence of organic matter. Biochemists believe that phenol and its derivatives act by coagulating proteins, especially in the cell membrane. Phenol is expensive, has a pungent odor, and is caustic to the skin, and therefore the role of phenol as an antiseptic has diminished.

However, phenol derivatives called *cresols* have greater germicidal activity and lower toxicity than the parent compound. Mixtures of ortho, meta, and para-cresol (creosote) are used commercially, as wood preservatives for railroad tics, fence posts, and telephone poles. Combinations of two phenol molecules called bisphenols are prominent in modern disinfection and antisepsis. Orthophenylphenol, for example, is used in Lysol, Osyl, Staphene, and Amphyl. Another bisphenol, hexachlorophene, was used extensively such products as toothpaste (Ipana), underarm deodorant (Mum), and bath soap (Dial). One product, pHisoHex, combined hexachlorophene with a pH-balanced detergent cream.

Pediatricians recommended it to retard staphylococcal infections of the scalp and umbilical stump, and for general cleansing of the newborn. However, study indicated that excessive amounts could be absorbed through the skin and cause neurological damage in newborn rats and monkeys, and hexachlorophene was subsequently removed from over the-counter products. PHisoHex is still available, but only by prescription.

One important bisphenol relative is *chlorhexidine*. This compound was approved by the Food and Drug Administration (FDA) for use as a surgical scrub, hand wash, and superficial skin wound cleanser. A 4 percent chlorhexidine solution in isopropyl alcohol is commercially available as Hibiclens. The chemical is believed to act on the cell membrane of Gram-positive and Gram-negative bacteria. Chlorhexidine is gradually assuming the widespread use once enjoyed by hexachlorophene. It does not appear to be absorbed through the skin. However, there is evidence that bacteria may grow within it. Another phenol derivative, *hexylresorcinol,* is used in a mouthwash and topical antiseptic (ST37) and in throat lozenges (Sucrets). It has the added advantage of reducing surface tension, thereby loosening bacteria from the tissue and allowing greater penetration of the germicidal agent.

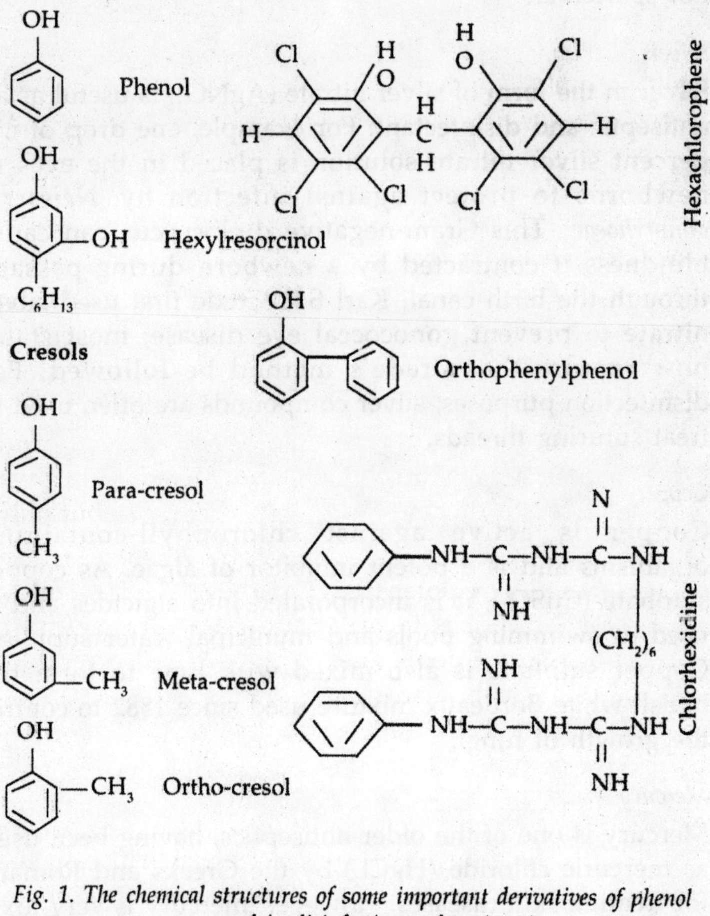

Fig. 1. The chemical structures of some important derivatives of phenol used in disinfection and antisepsis

Heavy metals

The elements such as mercury, silver, are called heavy metals because of their large atomic weights and complex electron configurations. Heavy metals are very reactive with proteins, particularly at sulfhydryl groups (—SH), and are believed to bind protein molecules together by forming bridges between them. Because the proteins arc often enzymes, the cellular metabolism is inhibited, and the microorganism dies. However, the heavy metals are not sporicidal.

Silver

Silver in the form of silver nitrate ($AgNO_3$) is useful as an antiseptic and disinfectant. For example, one drop of a 1 percent silver nitrate solution is placed in the eyes of newborns to protect against infection by *Neisseria gonorrhoeae.* This Gram-negative diplococcus can cause blindness if contracted by a newborn during passage through the birth canal. Karl S.F. Crede first used silver nitrate to prevent gonococcal eye disease; most states now require that Crede's method be followed. For disinfection purposes, silver compounds are often used to treat suturing threads.

Copper

Copper is active against chlorophyll-containing organisms and is a potent inhibitor of algae. As copper sulphate ($CuSO_4$), it is incorporated into algicides and is used in swimming pools and municipal water supplies. Copper sulphate is also mixed with lime to form the bluish white Bordeaux mixture used since 1882 to control the growth of fungi.

Mercury

Mercury is one of the older antiseptics, having been used as mercuric chloride ($HgCl_2$) by the Greeks and Romans for treating skin diseases. However, mercury is very toxic to the host, and the antimicrobial activity of mercury is reduced when other organic matter is present. In certain products such as Mercurochrome, Merthiolate, and Metaphen, mercury is combined with carrier compounds and is less toxic when applied to the skin, especially after surgical incisions. Merthiolate is also used as a preservative in vaccines and sera.

Alcohols

Alcohols are are valuable disinfectants for medical instruments. For practical use, the preferred alcohol is ethyl alcohol. Ethyl alcohol is active against vegetative bacterial cells, including the tubercle bacillus, but it has no effect on spores. It denatures proteins and dissolves lipids, an action that may lead to cell-membrane disintegration. Ethyl alcohol also is a strong dehydrating agent. Because ethyl alcohol reacts readily with any organic matter, medical instruments and thermometers must be thoroughly cleaned before exposure.

Usually, a 50 to 80 percent alcohol solution is recommended since water prevents rapid evaporation and assists penetration into the tissues. A 10-minute immersion in 70 percent ethyl alcohol is generally sufficient to disinfect a thermometer or delicate instrument. Alcohol is used to preserve cosmetics, and to treat skin before a venipuncture or injection. It mechanically removes bacteria from the skin and dissolves lipids to clean the skin. *Isopropyl alcohol*, commonly known as rubbing alcohol, is an effective substitute for ethyl alcohol, but methyl alcohol is toxic to the tissues.

Formaldehyde

Formaldehyde is a gas at high temperatures and a solid at room temperatures. When 37 grams of the solid are suspended in 100 ml of water, a solution called formalin results. For over a century, Formalin has been used in embalming fluid for anatomical specimens and by morticians. In microbiology, formalin is utilized for inactivating viruses in certain vaccines and producing toxoids from toxins. In the gaseous form, formaldehyde is expelled into a closed chamber where it is a sterilant for surgical equipment, hospital gowns, and medical instruments. However, penetration is poor, and the surface must be exposed to the gas for up to 12 hours for effective sterilization.

A mixture of formaldehyde and 70 percent isopropyl alcohol has been found to reduce to 3 hours the time necessary for spore destruction. In industry, formaldehyde is used to sterilize soil and disinfect imported fabrics. Many allergic individuals develop a contact dermatitis to this compound. Formaldehyde is an alkylating agent. It reacts with amino and hydroxyl groups of nucleic acids and proteins, and with carboxyl and sulfhydryl groups in proteins by inserting a small carbon fragment (an alkyl group) and forming bridges between the groups. This changes the structures of the molecules and interferes with an organism's chemistry, thereby leading to death.

Ethylene oxide

Development of plastics for use in microbiology required a suitable method for sterilizing these heat-sensitive materials. Research scientists discovered the antimicrobial abilities of ethylene oxide (EtO) and essentially made the plastic Petri dish and plastic syringe possible. Ethylene oxide is a small molecule with excellent penetration capacity and sporicidal ability.

However, it is both toxic and highly explosive. Its explosiveness is reduced by mixture with freon in Cryoxide or carbon dioxide in Carboxide, but its toxicity remains a problem for those who work with it. The gas is released into a tightly scaled chamber where it circulates for up to 4 hours with carefully controlled humidity. The chamber must then be flushed with inert gas for 8 to 12 hours to ensure that all traces of' EtO are removed, otherwise the chemical will cause "cold burns" on contact with the skin. Ethylene oxide has been employed to sterilize paper, leather, wood, metal, and rubber products, as well as plastics. In the hospital it is used to sterilize catheters, artificial heart valves, heart-lung machine components, and optical equipment.

The National Aeronautics and Space Administration (NASA) uses the gas for sterilization of interplanetary space capsules. Ethylene oxide chambers have become chemical counterparts of autoclaves for sterilization procedures. Often they are called Gas Autoclaves. A closely related gas, beta-propiolactone (BPL), is less explosive than ethylene oxide, but its penetrating capacity is more limited. As a liquid it is used to sterilize vaccines, sera, and surgical ligatures. However scientists have shown it to be carcinogenic, and BPL is therefore used only under restricted circumstances.

Glutaraldehyde

Glutaraldehyde is one of the most effective chemical liquids for sterilization purposes. This small molecule destroys vegetative cells within 10 to 30 minutes and spores in 10 hours. Glutaraldehyde is an alkylating agent, usually employed as a 2 percent solution. Because the activity of glutaraldehyde is not greatly reduced by organic matter, the chemical is recommended for use on surgical instruments where residual blood may be present.

In addition, glutaraldehyde does not damage delicate objects, and therefore it can be used to sterilize-optical equipment such as the optic fiber endoscopes used for arthroscopic surgery. Its disadvantages include the irritating fumes that it liberates and the need to rinse instruments thoroughly in sterile water. Automatic machines that incorporate glutaraldehyde are currently becoming more available.

Hydrogen peroxide

Hydrogen peroxide (H_2O_2) is used as a rinse in wounds, scrapes, and abrasions. The area foams and effervesces as catalase in the tissue breaks clown hydrogen peroxide to oxygen and water. This action assists mechanical removal of' microorganisms. Anaerobic bacteria may be inhibited by the sudden release of oxygen gas. The hydroxyl ions from water may also attract hydrogen ions away from certain proteins and cause a restructuring of the molecules. New forms of stabilized H_2O_2 such as Super D Hydrogen Peroxide contain no trace of' catalase and therefore do not decompose spontaneously. Such inanimate materials as soft contact lenses, utensils, heat sensitive plastics, and food processing equipment can be disinfected within 30 minutes. Sterilization is also achieved after 6 hours of exposure to a 6 percent solution. Microbiologists are currently researching the use of hydrogen peroxide for destroying bacteria in milk and milk products.

Soaps

Soaps are wetting agents, that is, they emulsify and solubilize particles clinging to a surface. The surface tension is also reduced by the soap. In addition, soaps remove skin oils, further reducing the surface tension and increasing the cleaning action.

Detergents

Detergents are synthetic chemicals acting as strong wetting agents and surface tension reducers. Since they are actively attracted to the phosphate group of cellular membranes, they also alter the membranes and encourage leakage from the cytoplasm. Certain *anionic* detergents yield negatively charged ions in solution.

The detergents are somewhat active against Gram-positive bacteria, but the negative charges of bacteria usually repel them and limit their use to common laundry products. Anionic detergents such as Triton W-30 and Duponol find value as iodophors. Other detergents are cationic. These derivatives of ammonium chloride contain four organic radicals in place of the four hydrogens, and at least one radical is a long-chain alkyl group. The positively charged ammonium group is counterbalanced by a negatively charged chloride ion. Such compounds are often called quaternary ammonium compounds or simply, quats. The compounds in cationic detergents have rather long, complex names, such as benzalkonium chloride in Zephiran and cetylpyridinium chloride in Ceepryn.

Other detergents are used in Phemerol and Diaparene. Cationic detergents are bacteriostatic on a broad range of bacteria, especially Gram-positive bacteria, and are relatively stable, with little odor. They are important sanitizers for industrial equipment and food utensils and are valuable in skin antisepsis because they dissolve the oils that surround organisms and remain active on the skin for long periods of time.

Dyes

A group of dyes called triphenylmethane dyes is also useful as antiseptics against species of Bacillus and Staphylococcus. The group include malachite green and crystal violet. Crystal violet has also been used traditionally as gentian violet for trench mouth and for *Candida albicans* infections such as thrush. Interference with cell wall construction appears to be the mode of activity. Antibiotics are currently the preferred method for treatment for both conditions. A second group of dyes, the acridine dyes, includes acriflavine and proflavine. Both dyes are used as antiseptics for staphylococcal infections in wounds.

Acids

Acid enhances the effects of disinfectants and antiseptics and makes them more soluble. Also, heat is a more potent sterilizing agent if acid conditions are present, because hydrolysis is increased. Acid is therefore a valuable adjunct to disinfection, albeit in an indirect way. Organic acids are particularly valuable as food preservatives. Lactic and acetic acids, for example, are important preservatives in sour foods such as cheeses, sauerkraut, and pickled products. Propionic acid is added to bakery products to keep bacterial populations low.

PART II
Neurobiology

Chapter 5

Synaptic Transmission

Our understanding about synaptic transmission has greatly increased. It is now apparent that the corticostriatal glutamatergic inputs produce rapid depolarization of striatal neurons via activation of ionotropic AMPA-type glutamate receptors. In addition, transmission is modulated by a number of presynaptic, G-protein-coupled receptors but, surprisingly, relatively little evidence of postsynaptic modulation has been observed. Corticostriatal synapses also express certain forms of plasticity, most notably short- and long- term synaptic depression (STD and LTD, respectively).

It appears that LTD may involve convergent actions of glutamate and dopamine. Striatal LTD may have important roles in information storage and motor set selection in the striatum. The striatum is a center for motor refinement and patterning, as well as storage of information regarding movement and other behaviours. One of the most important roles of the striatum is to facilitate the expression of intentional movements while suppressing unwanted movements. Several well-characterized disorders of movement and related cognition have their origins in damage to the striatum or its afferent connections. Included among these disorders are Huntington's disease and Parkinson's disease, as well as Tourette's syndrome. The normal functions of the striatum depend to a great extent on the complex interplay among different neurotransmitters and neuromodulators present in the striatal neurons themselves, as well as in the neurons that project to the striatum from the cortical and subcortical regions.

Recent years have witnessed a rapid expansion in our knowledge of the workings of synaptic connections within the striatum with the identification of the major chemical players in striatal circuitry and the neuronal sources of these chemicals. Forms of synaptic modulation, such as presynaptic control of neuro-transmitter release, are of importance for proper communication within this brain region. In addition, it is now becoming apparent that the striatum has mechanisms for the expression of long-lasting plasticity of synaptic transmission that help to endow this brain area with the capacity for long-term information storage. The cellular and molecular mechanisms contributing to striatal synaptic transmission, modulation, and plasticity are the target of much ongoing research.

In this review, we will focus on those mechanisms that operate in the neostriatum (caudate and putamen nuclei) and attempt to integrate this information into existing models of the function of striatal circuitry. The neostriatum contains two of the main entry nuclei into the subcortical circuitry known as the basal ganglia. The basal ganglia consist of subcortical nuclei in the telencephalon, diencephalon, and mesencephalon. The caudate and putamen nuclei develop from a telencephalic structure located in the floor of the lateral ventricles, whereas the globus pallidus is of diencephalic origin.

The subthalamic nucleus and substantia nigra develop from mesencephalic structures. These nuclei were first identified to be involved in the control of movement because disease states that affect the basal ganglia are characterized by involuntary movements or by a reduction in voluntary movement. The basal ganglia were originally identified as the major anatomical component of the extrapyramidal motor system, which was believed to act in parallel with but independently of the pyramidal, or corticospinal, motor system. It is now clear that these structures are also involved in cognitive and associative functions. Four major disease states are associated with the basal ganglia: Parkinson's disease, Huntington's disease, hemiballismus, and tardive dyskinesia.

Parkinson's disease is associated with the degeneration of dopamine-, serotonin-, and norepinephrine-containing neurons throughout the brain. The most prominent finding is a loss of dopamine in the brain and the associated degeneration and depigmentation of the substantia nigra that contains dopaminergic cells, which project to the striatum. The symptoms of Parkinson's disease include tremor at rest, increased muscle tone, akinesia, difficulty in initiating movement, and bradykinesia. Huntington's disease is characterized by the loss of acetylcholine- and γ-aminobutyric acid (GABA)-containing neurons within the striatum. The symptoms include dementia and chorea, or uncontrollable movements.

Hemiballismus occurs after damage to one of the subthalamic nuclei, usually as the result of an acute vascular accident, and is characterized by involuntary movements of the arm or leg on the contralateral side. Hemiballismus sufferers usually recover slowly over time. Tardive dyskinesia is caused by hypersensitivity of dopaminergic receptors, which can be the result of prolonged treatment with neuroleptics but can also be a complication of the treatment of Parkinson's disease. Investigation of the mechanisms of regulation of synaptic transmission at the corticostriatal synapse is of particular interest because of studies showing that the damage induced by excitotoxic concentrations of kainate, an agonist at glutamate receptors, mimics the damage seen in Huntington's disease.

Continual infusion of glutamate, the endogenous neuro transmitter at corticostriatal synapses, into the striatum could also elicit damage that resembled the pathological changes seen in Huntington's disease. This model has been further refined based on studies that indicated that agonists acting at N-methyl-D-aspartale (NMDA) type receptors reproduce the neuro-chemical and pathological changes more accurately than do kainate or glutamate. These findings suggest that drugs, which specifically regulate transmission at corticostriatal synapses, might be useful in treating Huntington's disease by preventing or retarding the degeneration of striatal neurons.

Anatomy of neostriatum

Neostriatum consists of the caudate and putamen nuclei. These nuclei receive topographically organized input from the entire cortex, as well as from the intralaminar nuclei of the thalamus. The input nuclei have topographically organized projections to the other nuclei of the basal ganglia. The striatum projects to both segments of the globus pallidus and to the substantia nigra pars reticulata. These efferent projections are GABAergic outputs that inhibit the function of other basal ganglia structures. The substantia nigra pars compacta provides a dopaminergic input to the neostriatum, whose function will be discussed later in this review. The dorsalraphe nucleus also provides a serotonergic input to the striatum.

The principal targets of output from the basal ganglia, and hence the systems that are modulated by striatal output, are nuclei of the thalamus. Input to these nuclei is relayed to the cortex. Thus, the function of the striatum and its targets in the basal ganglia is to process cortical output in order to provide feedback into the cortex to refine movements and sets of movements. The basal ganglia have been referred to as the extrapyramidal motor system, but it is now clear that they participate in a range of higher functions.

Almost all of the cerebral cortex projects to the striatum, and its inputs feed into an array of parallel pathways that regulate different aspects of motor function and cognition.

The *sensoly motor loop* is probably the best characterized pathway and consists of projections from somatosensory, primary motor, premotor, and supplementary motor cortices to the putamen that are then relayed back to the cortex though a complex series of interactions with other parts of the basal ganglia. This loop is the classical extrapyramidal loop for refinement of cortical motor output.

The *oculomotor loop* begins with projections from the posterior parietal cortex, the prefrontal cortex, and the frontal eye fields and projects to the body of the caudate nucleus and to other basal ganglia, eventually providing feedback to the cortex. Saccadic eye movements are at least partially controlled by this loop.

The *association loop* receives input from the posterior parietal, premotor, and dorsolateral prefrontal cortices. These afferent connections project to the head of the caudate nucleus that connects to the substantia nigra pars reticulata and the internal segment of the globus pallidus. Efferents from this loop project to the dorsolateral prefrontal cortex via the ventral anterior and dorsomedial nuclei of the thalamus. This circuitry is believed to be involved in some forms of information storage.

The fourth loop or limbic loop involves projections to the ventral striatum and head of the caudate nucleus, which project to the ventral pallidum, substantia nigra pars reticulata, and internal segment of the globus pallidus. Output of this loop is to the anterior cingulate and orbitofrontal cortices via the dorsomedial and the ventroanterior nucleus of the thalamus.

Neurotransmitters

Striatal cell population also includes interneurons: medium aspiny and giant aspiny neurons. Giant aspiny neurons are immunoreactive for choline acetyltransferase and make up 2% of the striatal cell population. They form cholinergic synapses with medium spiny neurons, as well as direct plasmalemmal associations. Based on electrophysiological data, giant aspiny neurons received monosynaptic cortical input. The giant aspiny neurons also receive dopaminergic input from the substantia nigra pars compacta. Another class of medium aspiny interneurons containing neuropeptide Y and somatostatin has also been shown to receive cortical and dopaminergic nigral afferent input. A third group consists of interneurons that are medium to large and have aspiny dendrites.

These neurons are immunoreactive for glutamate decarboxylase, which indicates that they use GABA as a neurotransmitter. An elegant study by Kawaguchi demonstrated that striatal interneurons can be subdivided into four classes based on intrinsic electrophysiological characteristics. This investigation included identification of these interneurons according to morphological characteristics and neurotransmitter phenotype. Staining of the striatum for acetylcholinesterase (AChE) reveals an uneven distribution with small AChE-poor areas embedded in an AChE-rich matrix. These small areas are referred to as striosomes.

Striosomes, first identified in human, monkey, and cat striatum, correspond to "patches" of high μ opioid receptor density in the rat striatum. Other neurotransmitters have also been found to have patchy distributions within the striatum. For example, regions of intense enkephalin immunoreactivity and regions of intense or low substance P immunoreactivity correspond to striosomes. An examination of immunoreactive cell bodies reveals that striosomes contain more cells that are positive for substance P, dynorphin, and neurotensin, whereas the matrix contains more cell bodies that are positive for enkephalin and somatostatin. Projections to the striatum show a specificity for striosome and matrix regions partially based on function. Sensorimotor areas project to the matrix, whereas limbic regions tend to project to striosomes. Dopaminergic projections from the substantia nigra are also split between the striosomes and the matrix. In addition to the inputs to the striatum being segregated between the striosomes and matrix, each compartment has distinct outputs.

The striosomes project primarily to the dopaminergic cells of the substantia nigra pars compacta. The matrix projects primarily to the substantia nigra pars reticulata and the globus pallidus. Projections from the matrix can be further subdivided based on the neurotransmitters expressed in each pathway. Cells that project to the internal segment of the globus pallidus and the substantia nigra pars reticulata contain GABA, dynorphin, and substance P. The striosomal cells projecting to the substantia nigra pars compacta contain a similar complement of neuro-transmitters. The cells that project to the external segment of the globus pallidus contain GABA, enkephalin, and neurotensin.

Physiological consequences of activation of the circuitry of basal ganglia can best be understood by assessing the impact of these circuits on cortical function.

The direct pathway, which involves GABAergic input from the striatum to the internal segment of the globus pallidus and the substantia nigra pars reticulata, inhibits the GABAergic outflow to the thalamus, leading to an increase in thalamic excitation of cortex. The indirect pathway, starts with GABAergic input from the striatum to the external segment of globus pallidus.

The output from the globus pallidus then inhibits the GABAergic output to the subthalamic nucleus. This leads to increased glutamatergic output to the globus pallidus and the substantia nigra pars reticulata, which will increase GABAergic outflow to the thalamus, resulting ultimately in a decrease in thalamic excitation of cortex. The dopaminergic projection from the substantia nigra pars compacta acts to stimulate the direct pathway and inhibit the indirect pathway.

Synaptic drive

Most striatal neurons show little activity in anesthetized animals or in the absence of movement. However, neurons will often show bouts of activity associated with movement. Based on neuronal action potential waveforms and firing patterns in vivo, two subtypes of neurons were described, type I and type II neurons. It is now apparent that the type II neurons correspond to the medium spiny principal striatal neurons, whereas the type I neurons are probably a mix of different types of interneurons. The different firing behaviour of these neurons can be attributed, in part, to differences in their passive membrane properties.

Intracellular and whole-cell recordings from medium spiny neurons show these neurons to have high resting membrane potentials and to exhibit no subthreshold excitatory conductance changes. Indeed, the only consistently observed subthreshold conductance is a-voltage-dependent K^+ current that would tend to prevent neurons from reaching firing threshold during small depolarizations. Medium spiny neurons are unlikely to fire in the absence of a strong, rapid, depolarizing stimulus such as a synaptic input.

On the other hand, some populations of interneurons appear to express active excitatory membrane conductances, including subthreshold calcium currents that promote cell firing. One of the important features of synaptic input to the striatum is that its volume must be sufficient to activate medium spiny neurons when necessary and to modulate activity if subtle alterations in firing rates are needed.

Corticostriatal synaptic transmission

The strong synaptic activation necessary to drive striatal medium spiny neurons appears to be provided by the cortical inputs. These inputs form glutamatergic synapses on the dendrites of the medium spiny neurons, which produce excitatory postsynaptic responses when activated. Excitatory glutamatergic synapses appear to be present on striatal interneurons. These synapses are not as well characterized, and thus modulation and plasticity at these synapses awaits further study.

However, improved techniques for identifying these neurons for intracellular recording should improve our understanding of these issues. Characterization of synapses on striatal neurons has benefited greatly from electrophysiological studies using *in vitro* preparations. Using striatal slices and recording extracellular field potentials, Yamamoto found that electrical stimulation within the striatum produced a response with two prominent negativities.

The first negativity (N1) represents direct stimulation of axons And intrinsic cells, whereas the second negativity (N2) represents the synaptically mediated field potential. Under conditions that reduce synaptic transmission, such as decreased extracellular Ca^{2+} or increased extracellular Mg^{2+} it was found that N2 was reduced or abolished but that N1 was unaffected. Furthermore, both N1 and N2 are reversibly blocked by the local anesthetic procaine, which blocks the voltage-gated Na^{+} channels responsible for the generation of action potentials. Intracellular recordings have demonstrated that intrastriatal stimulation elicits monosynaptic excitatory post synaptic potentials (EPSPs).

These EPSPs are abolished with reduction of extracellular Ca^{2+} and increased Mg^{2+} and in the presence of cobalt or TTX, indicating that they are synaptically driven. The EPSPs elicited by intrastriatal stimulation are rarely contaminated by inhibitory postsynaptic potentials (IPSPs) or by depolarizing IPSPs, whereas others report that application of GABAA antagonists bicuculline or picrotoxin, or injection of Cl^{-} into the postsynaptic cell increases the amplitude of the EPSP, indicating that the EPSPs are contaminated with GABAergic IPSPs.

Others have shown that at resting membrane potentials the GABAergic postsynaptic potential is depolarizing and application of bicuculline or picrotoxin actually reduces the EPSP. Seabrook and colleagues found that the ability to elicit an EPSP with a GABAergic component was dependent on the orientation of the stimulus, with dorsal stimulation eliciting an EPSP without a GABAergic component, whereas ventral stimulation elicited an EPSP with a GABAergic component. Jiang and North found that EPSPs containing a GABAergic component could be elicited by intrastriatal stimulation but not by stimulation in the corpus callosum or the external capsule. Experiments with striatal slices have generally confirmed that activation of cortical afferents, either by direct cortical stimulation or by intrastriatal stimulation, produces fast glutamatergic transmission.

Takagi and Yamamoto found that curare did not reduce N2 but that activation of muscarinic receptors by acetylcholine and carbamylcholine did do so. Other investigators later demonstrated that N2 was reduced by excitatory amino antagonists. Nicotinic and muscarinic antagonists, as well as dopaminergic antagonists, did not affect N2 but that the, excitatory amino acid antagonists, g-D-glutamylglycine and (±)-cis-2,3-piperidine dicarboxylate, reduced, or abolished N2. This was supported by the findings of Malenka and

Kocsis that neither nicotinic nor dopaminergic antagonists reduced N2 but that the excitatory amino antagonists kynurenate, gamma-E-glutamylglycine, and piperidine dicarboxylate reduced or abolished N2. Kynurenate reduced N2 in a dose-dependent manner at all stimulus intensities. Stimulation in corpus callosum, in which the only striatal afferents are crossed corticostriatal fibers, elicited a kynurenate sensitive N2 similar to that elicited with intrastriatal stimulation. This suggested that the excitatory response was mediated by stimulation of glutamatergic corticostriatal fibers within the striatum. The EPSP elicited by intrastriatal stimulation also appears to arise from activation of glutamatergic synapses. EPSPs evoked by intrastriatal stimulation could be reduced in a dose dependent manner by the nonspecific excitatory amino acid antagonists kynurenate and gamma-D-glutamylaminomethyl sulphonate (GAMS). They also demonstrated that (D.L)-2-amino-7-phosphonoheptanoic acid [(D,L) AP7], an antagonist for NMDA type glutamate receptors, did not affect the EPSP under control conditions, but when the cell membrane was depolarized above -50 mV or if the extracellular Mg^{2+} was removed, AP7 partially reduced the EPSP.

Nicotinic antagonists tubocurarine and hexamethonium did not reduce the EPSP. Other investigators have demonstrated that EPSPs elicited by electrical stimulation in the corpus callosum, or deep layers of cortex are reduced in a dose-dependent manner by kynurenate. These studies support the conclusion that stimulus-evoked excitatory transmission in vitro striatal slice preparations results mainly from stimulation of corticostriatal afferent fibers and is mediated by glutamate. The role of NMDA and non-NMDA postsynaptic ionotropic glutamate receptors has been examined extensively in striatum. With the development of selective non-NMDA antagonists, investigators have demonstrated that the EPSPs elicited by both intrastriatal and callosal stimulation have a non-NMDA and an NMDA component, although the non-NMDA component predominates under conditions designed to approximate the normal environment of the brain.

The role of NMDA receptors at these synapses is not well understood. However, medium spiny neurons do express appreciable numbers of these receptors and NMDA receptors- dependent responses can be observed in striatal slices under conditions of reduced extracellular Mg^{2+}. The literature concerning the involvement of NMDA receptors in striatal function and neurochemistry is extensive and in particular, neurochemical studies have suggested interactions between NMDA receptors and dopaminergic transmission. N2 and the EPSP elicited by intrastriatal stimulation were mediated by acetylcholine. Misgeld and colleagues reported that N2 was blocked by the nicotinic antagonists tubocurarine and mecamylamine and concluded that N2 was mediated by acetylcholine acting on nicotinic receptors and also that the source of acetylcholine was striatal interneurons. In addition, the EPSPs elicited by intrastriatal stimulation were also reported to be mediated by acetylcholine acting on nicotinic receptors.

Dopaminergic nigrostriatal synaptic transmission

Importance of the dopaminergic input is well established, we do not yet have a clear understanding of the cellular physiology involved in the actions of dopamine. Dopamine is presumed to act as a neuromodulator because monoamines generally perform modulatory functions in the brain. The known dopamine receptor subtypes are G protein-coupled receptors. Activation of D1-like dopamine receptors stimulates adenylate cyclase, increasing the intracellular concentration of cAMP.

However, the cellular consequences of this increase in second messenger formation are unclear. D2-like receptors activation inhibits adenylate cyclase activity. Both neurochemical effects are clearly mediated through activation of G proteins. It is likely that a Gs-like G protein is involved in the cyclase stimulation by D1-like receptors, whereas Gi or a related G-protein mediates the cyclase inhibitory effects of D2-like receptors.

These observations further support the idea that dopamine has a modulatory role in striatum since G protein-coupled receptors are usually involved in neuromodulation. However, the exact nature of this modulatory influence is unknown. Dopamine can alter the physiology of striatal neurons in different ways. Dopamine application produces both excitatory and inhibitory actions on single striatal neurons when extracellular and intracellular electrophysiological recordings have been performed.

Studies using intracellular recording, whole-cell recording, and recording from single channels in membrane patches have revealed it few potential mechanisms by which dopamine alters striatal neurophysiology. For example, dopamine can activate single K^+ channels in membrane patches of isolated striatal neurons, and involvement of D2-like receptors in this phenomenon has been suggested. A similar action of dopamine has also been demonstrated in neurons of the substantia nigra. However, this inhibitory action of dopamine has not been reported in whole cell recordings or sharp-electrode intracellular recordings from striatal neurons but has been observed at the whole-cell level in substantia nigra.

It is possible that the K^+ channels activated by dopamine in striatum are different from those activated in other brain regions. Dopamine can also both inhibit and potentiate the function of voltage-gated Na^+ and Ca^{2+} channels. The direction of the response appears to depend on the type of receptor activated and the molecular target affected. D1-like receptor effects on Na^+ channels appear to involve cyclic AMP and PKA-mediated protein phosphorylation, whereas effects on Ca^{2+} channels appear to involve activation of both PKA and phosphatase 1.

On the other hand, D2-like receptors produce potentiation of Na^+ channel function via a separate mechanism. However, the alterations in action potential frequency or threshold resulting from this effect have not been elucidated, although dopamine effects on cell firing have been reported in experiments on striatal slices and have been attributed to activation of D1 receptors.

Dopamine alters glutamatergic corticostriatal transmission. Levine and coworkers have observed that dopamine has differential effects on responses mediated by different ionotropic glutamate receptors. Calabresi and coworkers have also examined dopamine effects on corticostriatal transmission in striatal slices from animals after chronic neuroleptic treatment or dopaminergic denervation in comparison to slices from untreated animals. These investigators have reported that dopamine only inhibits transmission after chronic neuroleptic treatment of striatum. This may be attributable to upregulation of dopamine receptors following neuroleptic treatment.

Subtypes of dopamine receptors involved in this synaptic modulatory effects are believed to be D2-like. Despite these indications of the physiological effects of dopamine, it is still not clear how this neurotransmitter has such a large influence on striatal function. In particular, there is little information about the effects of activation of nigral afferents on the excitability of striatal neurons and corticostriatal transmission.

Presynaptic regulation

One of the most striking features of corticostriatal synaptic transmission is strong presynaptic modulation by several neurotransmitters. Presynaptic modulation has been described at a number of synapses in the central nervous system, most notably in the spinal cord and hippocampus. Presynaptic inhibitory effects of a number of neurotransmitters at corticostriatal synapses have been documented. These transmitters include acetylcholine, adenosine, GABA, glutamate, and endogenous opiates.

An exhaustive investigation of the subtypes of receptors that mediate presynaptic inhibition at this synapse has not been undertaken, and thus it is possible that this list will be augmented in the future. Evidence that activation of these receptors inhibits transmission at a presynaptic locus includes the fact that activation of the receptors does not have detectable postsynaptic effects and that postsynaptic responses produced by application of glutamate or glutamate analogs are not inhibited by activation of these receptors. The presynaptic receptors characterized to date appear to be G protein-coupled receptors of subtypes that are known to inhibit adenylate cyclase.

However, it is not clear that cyclase inhibition plays any role in presynaptic inhibition by these receptors. It is currently thought that these inhibitory receptors modulate transmitter release via inhibition of voltage-gated calcium channels or more direct effects on vesicular secretion.

Another interesting aspect of the physiological roles of G protein-coupled receptors in neostriatum is the relative lack of evidence for postsynaptic actions. G protein-coupled receptors of the types that inhibit adenylate cyclase often activate K^+ channels in central and peripheral neurons. However, this inhibitory response has not been observed in striatal medium spiny neurons upon activation of GABA, adenosine, opiate, or metabotropic glutamate receptors. The reason for this lack of activation of inward rectification is not clear. It is possible that striatal medium spiny neurons do not express an inwardly rectifying K^+ channel of the type that is usually activated by G protein-coupled receptors.

Indeed, there is very little evidence of inward rectification when recording from these neurons in the absence or presence of added neurotransmitter. However, this does not constitute strong evidence for lack of inward K^+ rectification because this current is often quite small in the absence of receptor agonist in hippocampal neurons.

The possibility that functional GIRK channels and G proteins exist in striatal medium spiny neurons needs to be examined more extensively, perhaps by examining the effect of direct G protein activation on ion current in medium spiny neurons. However, it is clear at this time that several modulatory transmitters have little influence on the postsynaptic properties of medium spiny neurons. The reason for strong presynaptic control of corticostriatal synaptic transmission is not clear.

However, it is possible that presynaptic inhibition serves as a selectivity filter for the transduction of cortical input into striatal input. This means that when more than one set of cortical afferents impinges on a given set of striatal medium spiny neurons, it may be the relative level of presynaptic modulation, as well as the rate of afferent firing and the number of synapses, that determines the likelihood of activating the postsynaptic neuron. This would be an effective means of repressing less important afferent input at a time when a particular motor set requires activation.

One can easily imagine that highly active glutamatergic cortical afferents may inhibit afferents at other synapses in the striatum via activation of presynaptic metabotropic glutamate receptors (mGluRs). Similarly, release of adenosine during intense afferent activation may shut down particular lines of cortical input into the striatum. Thus, presynaptic inhibition at corticostriatal synapses is probably of great importance in selecting appropriate input-output relationships between cortex and striatum. The role of dopamine in the process of presynaptic filtering would be prominent, considering the large dopaminergic projections into the striatum. However, the role of presynaptic dopamine receptors at corticostriatal afferents is not well understood.

Dopamine D2 receptors have been localized to presynaptic terminals in striatum, including glutamatergic cortical afferent terminals. Evidence from neurochemical studies also indicates that dopamine can inhibit the accumulation of extracellular glutamate. Direct and indirect dopamine receptor activation decreases the excitability of corticostriatal afferent terminals. However, no consistent effects of dopamine application on corticostriatal transmission have been observed in electrophysiological experiments in striatal slices from drug-naive rats.

It is clear that presynaptic inhibition by dopamine D2 receptors increases following chronic neuroleptic treatment or dopaminergic denervation in animal models. This observation suggests a potential mechanism by which overcompensation for neuroleptic actions may disrupt nigral control of striatal output. However, it is not yet clear whether the presynaptic effect of dopamine is a key factor in controlling normal striatal function or is merely a manifestation of drug-induced striatal pathology.

It may be important in future investigations to examine dopamine effects on individual corticostriatal synapses using analysis of miniature synaptic currents. It is possible that selected subgroups of cortical afferents are more strongly affected by dopamine than are others. If dopamine-sensitive synapses are identified then it might be possible to determine whether these synapses are made by afferents from particular cortical regions or whether they impinge on selected portions of the dendritic arbors of medium spiny neurons.

Striatal medium spiny neurons are especially susceptible to excitoxicity induced by glutamate and glutamate analogues. The first protective action during neuronal insult or a hypoxic state would be to prevent or curtail exposure to glutamate by shutting down stimulated release of the neurotransmitter. In this regard, the finding that adenosine accumulates extracellularly during the early stages of hypoxia is especially interesting. A surge in extracellular glutamate also occurs during and after ischemia and hypoxia. Glutamate may feed back onto mGluRs on corticostriatal synapses to inhibit further release and thus curtail the length of the glutamate surge. However, it should be noted that some of the glutamate release may be nonsynaptic and thus may not be subject to inhibition by G protein-coupled receptors.

Relationship between excitotoxicity processes and the genetic defect that appears to underlie the disease is by no means clear. It will be interesting to determine whether glutamate release and regulation of release by presynaptic receptors is involved in any excitotoxic underpinnings of this disease. Compounds present at corticostriatal synapses and with actions on presynaptic receptors may be useful therapeutic agents for the treatment of neurological disorders involving the striatum.

Depression at corticostriatal synapses

The modulation of corticostriatal transmission via activation of presynaptic receptors is likely to lead to changes in transmission that last from seconds to minutes. Indeed, adenosine A1 receptor-mediated synaptic depression persists for several minutes following high frequency afferent activation. Recently, synaptic transmission at corticostriatal synapses has been shown to exhibit longer-lasting changes in synaptic efficacy. These more durable forms of synaptic plasticity could play a role in storage of information relevant to the performance of motor programs or simple operant tasks. A persistent depression of corticostriatal synaptic transmission occurs following high frequency stimulation in the cortex or within the striatum itself.

The intensity of stimulation that evokes this long-term depression (LTD) is higher than that needed to evoke shorter-lasting adenosine receptor-mediated synaptic depression. Studies by Calabresi and colleagues indicate that the need for strong synaptic activation in LTD initiation may be related to the need to achieve strong postsynaptic depolarization and activation of L-type Ca^{2+} channels. These investigators have demonstrated that initiation of LTD is prevented if the postsynaptic membrane potential is hyperpolarized during high frequency synaptic activation. Application of Ca^{2+} chelators such as EGTA of BAPTA to the interior of striatal neurons and treatment with dihydropyridine L-type Ca^{2+} channel blockers also prevents LTD initiation. Together, the last two pieces of evidence indicate that an increase in intracellular Ca^{2+} via activation of L-type channels may be a crucial step in the processes linking high-frequency stimulation to synaptic depression.

Postsynaptic cell depolarization during high frequency synaptic activation has thus been proposed to lead to activation of L-type Ca^{2+} channels, which in turn, results in a sharp rise in intracellular Ca^{2+}. The steps linking the rise in Ca^{2+} concentration to the decrease in synaptic efficacy are not yet clear, but activation of Ca^{2+} dependent protein kinases may play a key role. The strong depolarization needed to activate Ca^{2+} channels during LTD initiation may result from persistent release of glutamate or release of another neuro-transmitter. Pharmacological studies suggest that both may occur. It has been demonstrated that LTD is blocked if high frequency stimulation is given during application of the mGluR antagonist AP3 or during application of either D1-like or D2-like dopamine receptor antagonists, and that LTD is blocked by depletion of dopamine.

These observations implicate all three receptor subtypes in LTD initiation. The observations with dopamine receptor antagonists suggest that release of endogenous dopamine takes place during high-frequency stimulation. This intriguing finding may herald a new role for striatal dopamine: activation of synaptic plasticity that might alter striatal and, therefore, cortical motor output. However, a number of issues remain to be resolved before an accurate picture emerges of the role of glutamatergic and dopaminergic transmission in initiating striatal LTD.

For example, it is not clear if activation of ionotropic glutamate receptors is needed to provide some of the postsynaptic depolarization that is crucial to LTD initiation. LTD can be initiated in the presence of an NMDA receptor antagonist. Thus, if any ionotropic receptor is involved, it is most likely an AMPA/ kainate type of receptor.

Mechanisms underlying dopamine release during high frequency stimulation are also unclear. In the experiments in which dopamine receptor blockade prevented LTD initiation, high frequency stimulation was applied to the deep layers of cortex. Thus, one would not expect to elicit electrical stimulation and direct activation of nigral dopaminergic afferents that are only present in the striatal portion of this slice preparation. It is possible that some feedback between striatal postsynaptic neurons and dopaminergic afferents or signalling between glutamatergic and dopaminergic synaptic terminals occurs during high frequency stimulation. There is some evidence for the presence of glutamate receptors on striatal dopaminergic terminals, which can stimulate dopamine release as mentioned above.

However, indirect mechanisms linking GluR activation to dopamine release (for example, $[K^+]o$ accumulation leading to terminal depolarization) have not been ruled out. The mechanism by which activation of dopamine receptors leads to LTD initiation is not well understood. As mentioned earlier, the physiological consequences of dopamine receptor activation in striatum are not well characterized. Thus, there may be changes in postsynaptic or presynaptic properties that contribute to the initiation of LTD that have not yet been explored. The coupling of different dopamine receptor subtypes to adenylate cyclase produces opposing effects—D1-like receptors stimulate cyclase activity, whereas D2-like receptors inhibit activity.

It is unlikely that these actions play a role in LTD because it is difficult to imagine a synergistic action of these two receptors on adenylate cyclase. Temporally coordinated changes in cAMP formation may be needed for LTD. It has also been suggested that stimulation of arachidonic acid formation by dopamine plays a role in LTD initiation. Activation of both D and D2-like receptors leads to synergistic increases in the formation a release of arachidonic acid. It is possible that the fatty acid serves as an extracellular messenger coupling postsynaptic activation to presynaptic processes that might be involved in LTD. This scenario is similar to that proposed by some groups to underline hippocampal LTP.

Arachidonic acid inhibits the firing of striatal neurons in culture, and this has been suggested to be important for LTD. However, LTD induced by high frequency afferent stimulation does not involve changes in striatal neuron firing or excitability, and thus it is unlikely that the effects of arachidonic acid on striatal neurons are linked to LTD at corticostriatal synapses. The role of mGIuR activation in initiation of LTD also remains to h clarified. The mGluR antagonist that blocks LTD, AP3, is a weak antagonist that may only be effective at group I mGluRs. This subclass of mGluRs is known to stimulate hydrolysis of phosphatidylinositol (PI) and increases in intracellular Ca^{2+} and to depolarize certain types of neurons.

All of these intracellular signals could contribute to initiation of LTD. Stimulation of PI hydrolysis and increases in intracellular Ca^{2+} would likely stimulate Ca^{2+}-dependent protein kinases and other enzymes whose function may be necessary for initiating LTD. Specific tests for the involvement of this signalling pathway in the molecular events linking mGluR activation to LTD are needed. It is also possible that mGluR activation contributes to the depolarization of striatal neurons, which is necessary for initiation of LTD.

However, application of the mGluR agonist trans-amino cyclopentane dicarboxylic acid, which should activate all mGluR subtypes, does not produce strong depolarization of striatal neurons. The membrane potential and other electro-physiological characteristics of striatal neurons are not altered in the majority of striatal neurons exposed to mGluR agonists. Thus, it appears to be unlikely that mGluR activation contributes much to the depolarization that is necessary for the onset of LTD. Initiation of striatal LTD is also blocked by acute or chronic exposure to lithium. This is an interesting observation because therapeutic use of lithium has been reported to produce effects on motor function.

The mechanism underlying this effect of lithium is not currently known. However, lithium can alter PI hydrolysis, and this action may interfere with intracellular signals activated by increases in Ca2+ during the initiation of LTD. LTD expression is synapse specific, which means that, synapses that are not activated at high frequencies do not exhibit LTD even when the synapse is present on the same postsynaptic neuron as other synapses exhibiting LTD. This synapse specificity is reminiscent of long lasting forms of synaptic plasticity in the hippocampus and suggests that maintenance of LTD involves highly spatially localized changes in synaptic biochemistry, morphology, or function. This is an important property of any form of synaptic plasticity because it serves to alter the balance of information coming from different afferent inputs without changing the response of the postsynaptic neuron to all afferent inputs. An in vivo study suggested that LTD involves a change in the excitability of presynaptic afferent fibers.

However, it is not yet clear whether LTD observed after afferent stimulation in vivo is the same as that observed in striatal slices. None of the observations made to date in the slice preparation are incompatible with a change in presynaptic excitability. The postsynaptic locus of control of LTD initiation complicates explanations of the mechanisms leading to a presynaptic change. However, several hypotheses that might explain such a mechanism have been put forward by investigators examining hippocampal LTP.

Thus, a change in afferent excitability remains a viable explanation for the expression of striatal LTD. The excitability and electrophysiological characteristics of striatal medium spiny neurons are not altered as a result of LTD expression. For example, the resting membrane potential and input resistance of these neurons and the firing of action potentials in response to depolarization does not change when LTD is expressed. These observations strongly suggest that LTD is not the result of a decrease in the ability of the postsynaptic neuron to respond to strongly depolarizing stimuli.

However, the lack of changes in these postsynaptic responses should not be interpreted as a lack of involvement of any postsynaptic contribution to the maintenance of LTD. Subtle changes in neuronal electrophysiological properties at the synapse may not be detected when the physiological responses to current injected into the cell body are examined.

Furthermore, the postsynaptic response to application of exogenous glutamate has not been studied. It should be noted, however, that this latter experiment may not provide much information because it is the change in response to glutamate at depressed synapses and not on the neuron as a whole that is of interest. Despite these caveats, the observations made to date do suggest that LTD does not result in a generalized change in postsynaptic neuron function. The development of techniques for proper measurement and analysis of unitary synaptic events at corticostriatal synapses is likely to be of considerable importance in assessing possible presynaptic and postsynaptic contributions to LTD.

Changes in the amplitude of miniature synaptic currents can provide information about the status of postsynaptic responses generated at the synapse itself and thus represent a better measure of synapse-specific postsynaptic changes. Analysis of unitary synaptic currents in striatal slices may be possible if initial technical difficulties can be overcome. It is possible to record these responses in a corticostriatal coculture system developed in our laboratory. Whether synapses in this preparation are capable of expressing LTD remains to be determined. Determining the presynaptic or postsynaptic locus of the physiological changes crucial for expression of striatal LTD may prove to be as challenging as determining the locus of changes important for hippocampal LTP.

However, investigators interested in addressing this question can draw on the substantial literature on hippocampal UP when designing proper experiments to test key hypotheses. For example, it has become clear that certain assumptions made from quantal analysis at synapses in the peripheral nervous system may not be valid for central synapses. Thus, recent evidence suggests that synapses in the hippocampus may appear to be "silent" if they contain predominantly NMDA type glutamate receptors. This results in a situation in which a decrease in "failures," which is usually interpreted as a presynaptic change, could result from a postsynaptic increase in non-NMDA receptor function.

In addition, quantal analysis at some hippocampal synapses has been hampered by lack of a good fit to traditional models, as well as by the problems associated with detecting small changes in quantal parameters at synapses distant from the somatic recording site. This literature should be studied before a similar analysis of striatal LTD is undertaken.

Finding the locus of expression of striatal LTD may benefit from the development of model systems in which postsynaptic function can be isolated. Such models have been used successfully in the study of LTD at parallel fiber synapses onto cerebellar Purkinje neurons. At these synapses, LTD is almost certainly expressed as a decrease in response to glutamate, probably involving a change in AMPA receptor function or number.

Changes in Purkinje neuron responses to AMPA application have been detected after treatments meant to mimic postsynaptic responses to paired synaptic activation in a culture system. The simplified LTD model has been extremely useful for assessing the role of different receptors and signalling pathways in cerebellar LTD. Striatal LTD appears to share some common features with cerebellar LTD, including dependence on postsynaptic control of initiation on activation of voltage-gated calcium channels, involvement of protein kinases, and involvement of AP3-sensitive metabotropic glutamate receptors.

It is possible that intracellular mechanisms underlying the two forms of LTD may be similar, and that primary cultures of striatal neurons may prove useful in elucidating postsynaptic mechanisms that contribute to striatal LTD. The role in LTD expression of intracellular signalling systems remains to be explored. Blockade of protein kinases inhibits LTD initiation. However, the effects of kinase inhibitors after the initiation of LTD have not yet been explored. Striatal medium spiny neurons are known to be rich in DARPP-32, a phosphatase inhibitor that is regulated by dopamine through cAMP.

It is possible that the intracellular cascade involved in LTD initiation and maintenance involves dopaminergic regulation of the function of this abundant protein. The role of changes in protein expression and morphological factors such as involvement of cytoskeletal elements or changes in synaptic structure have not been evaluated. It is tempting to speculate that LTD maintenance may involve changes in phosphorylation of proteins such as MAP2 that are subject to phosphorylation and could be involved in morphological changes at synapses. These factors have proven to be important components of expression of synaptic plasticity at other synapses in the brain, and thus it will be interesting to determine whether they play a prominent role in striatal LTD.

Long-term potentiation (LTP), the most widely studied example of a long duration change in synaptic efficacy, is observed at many synapses in he brain, and may be involved in information storage in the hippocampus and elsewhere. Calabresi and coworkers have demonstrated that corticostriatal synapses will undergo UP if high frequency afferent stimulation is given under conditions of reduced extracellular Mg^{2+}.

This switch from LTD to UP appears to involve activation of NMDA-type glutamate receptors. It is difficult to activate these receptors at corticostriatal synapses in the slice preparation under physiological conditions. Lowering extracellular Mg^{2+} relieves the blockade of NMDA receptor-channels by this ion and leads to enhanced synaptic responses mediated by these receptors. Strong activation of these receptors leads to expression of LTP, presumably via mechanisms similar to those involved in expression of hippocampal LTP. However, it is not yet clear if this form of synaptic plasticity is ever expressed in the striatum under conditions more closely approximating the in vivo situation.

Enhancing NMDA receptor function would probably increase the likelihood of inducing LTP. As yet, treatments that specifically enhance the function of these receptors and that are likely to occur during normal brain function have not been found. It will be interesting to determine if modulatory neurotransmitters such as acetylcholine or neuropeptides or protein kinase activation can enhance NMDA receptor function and the likelihood of eliciting LTP. Identification of such mechanisms would increase confidence that corticostriatal synapses are capable of the sort of dual regulation seen in the hippocampus and cortex.

Plasticity at nigrostriatal dopaminergic synapses

Lack of direct electrophysiological evidence about the nature of dopaminergic nigrostriatal synaptic transmission in real time means that most of our understanding of plasticity at these synapses comes from pharmacologic, neurochemical, and behavioural studies. As mentioned earlier, chronic neuroleptic treatment leads to upregulation of the function of D2-like dopamine receptors. A similar effect of denervation of dopaminergic input to striatum has also been observed. These changes presumably lead to an increased inhibitory influence on corticostriatal transmission.

There is good evidence that presynaptic D2-like autoreceptors exist on the terminals of dopaminergic neurons. Thus, this supersensitivity to dopamine might also be expected to alter the strength of transmission at these synapses. Nigral dopaminergic neurons possess intrinsic membrane properties that cause them to exhibit spontaneous pacemaker firing activity. Thus, it is likely that these neurons are constantly transmitting modulatory information at high frequency. Alterations in the activities of these neurons usually take the form of changes in bursting frequency or bursting parameters. If plasticity at these synapses is initiated by changes in the frequency of afferent input, as is suspected for most CNS synapses, then plasticity at nigrostriatal synapses can be expected to be a very dynamic process.

Chapter 6

Ion Channels

Ion channels are multimeric structures, often pentomers of different subunits, that gate ions following subtle changes in a transition tertiary structure. Small changes in tertiary structure result transition from the closed to the open to the inactivation state. These transitions may require the movement of only short stretches of amino acids from a buried position in the multimeric molecular structure to the extracellular fluid or to other intramolecular domains that regulate the gating properties of the ion channel.

Ion channels show a sensitivity to alcohol and anesthetic potency that correlate with the hydrophobicity of the alcohol or anesthetic being studied. The major excitory and inhibitory amino acid receptors in the mammalian central nervous system are considered to be glutamate, γ-aminobutyric acid type A ($GABA_A$), and glycine receptors. These receptors are widely acknowledged to participate in neurotransmission, and they also form target substrates for phosphorylation.

The process of phosphorylation involves the transfer of a phosphate group(s) from adenosine triphosphate to one or more serine, threonine, or tyrosine residues, which are invariably found in an intracellular location within the receptor. Phosphorylation is an important means of receptor regulation since it represents a covalent modification of the receptor structure, which can have important implications for ion channel function. Fast synaptic transmission in the central nervous system (CNS) is believed to be mediated mainly by amino acid neurotransmitters that activate ligand-gated ion channels.

The major excitatory transmitter in the brain is thought to be L-glutamate whereas γ-aminobutyric acid (GABA) and glycine serve as the major inhibitory neurotransmitters in the brain and spinal cord, respectively. Activation of these receptors results in the rapid flux of ions through integral ion channels, which results in depolarization or hyperpolarization of the target neuron. Since receptors for these neurotransmitters play key roles in controlling neuronal activity, there is considerable interest in how these receptors are regulated.

One mechanism that has received considerable attention is the covalent modification of receptor structure via phosphorylation. Phosphorylation is a process by which the highly charged terminal phosphate group from adenosine triphosphate (ATP) is transferred to serine, threonine, or tyrosine residues within a substrate, thus modifying its structure and possibly its function. This covalent modification is catalysed by a series of enzymes known as protein kinases, whereas the cleavage of the phosphate group from the phosphoprotein is catalysed by phosphoprotein phosphatases.

The most commonly known class of protein kinases are the classical serine/threonine second messenger-dependent protein kinases, such as cAMP-dependent protein kinase (PKA). In addition, there are many serine/threonine protein kinases that are activated by signalling cascades, and there is an additional family of kinases that specifically phosphorylates tyrosine residues. The expression levels of many of these kinases are highest in the nervous system, suggesting an important role in neuronal function. In this review, we discuss experimental evidence that suggests amino acid neurotransmitter receptors can be modulated by phosphorylation and we indicate the possible relevance of this regulation to neuronal function. Significant interest in determining the minimal requirement for substrate recognition by protein kinases has led to the identification of "consensus sites" for phosphorylation within proteins.

Significant progress has recently been made in identifying consensus recognition sites for a range of serine/ threonine and tyrosine kinases utilizing orientated peptide libraries. Consensus sites usually take the form of a short linear sequence surrounding the site (s) of phosphorylation identifying the minimum set of amino acids required for substrate recognition. There are, however, significant limitations to this approach.

Many kinases show a very broad substrate specificity [for example, protein kinase C (PKC)], and factors such as secondary and tertiary structure can be of importance in substrate recognition. Therefore, the existence of an apparent consensus site for phosphorylation neither ensures that a protein can be phosphorylated nor is an accurate indicator of which kinase (s) is responsible for phosphorylation. At best, consensus sites serve only as a guide whose information must be confirmed experimentally. A further problem in the case of ligand-gated ion channel subunits is that the structure of these Proteins has not been elucidated at high resolution.

Therefore, predictions of membrane topology are largely based on hydropathy profiles derived from primary amino acid sequences. Because protein kinases and protein phosphatases are almost exclusively intracellular molecules, the accurate prediction of which residues are likely kinase substrates depends on defining the intracellular domains of receptor subunits. Using hydropathy profiles to determine intracellular and extracellular domains of receptor subunits has severe limitations as demonstrated by the conflicting evidence regarding the transmembrane topology of both the *N*-methyl-D-aspartate (NMDA) and the non-NMDA receptor subunits.

$GABA_A$ receptors

$GABA_A$ receptors are found throughout the CNS and are the sites of action of many important therapeutic drugs, including the benzodiazepines, barbiturates, anaesthetic steroids, and convulsant agents such as picrotoxin. Seminal cloning studies have revealed that $GABA_A$ receptors are members of a channel superfamily that includes nicotinic acetylcholine receptors (AChRs), serotonin ($5\text{-}HT_3$) receptors and glycine receptors (GlyRs).

Members of this family share many conserved features including a common proposed membrane topology consisting of a large glycosylated N-terminal extracellular domain, four transmembrane domains (TMs), and a major intracellular domain between TM3 and TM4. This predicted intracellular domain accounts for up to 20% of the total mass of a typical receptor subunit. Several $GABA_A$ receptor subunits have been identified, which can be divided into subunit classes based upon the deduced amino acid sequence homology: α (1-6), β(1-4), γ (1-4) and δ(1). In addition, two homologous cDNAs, $\rho 1$ and $\rho 2$, which are expressed in the retina and encode bicuculline-insensitive $GABA_C$ receptor subunits, have also been isolated. Typically there is 30-40% homology between subunit classes and 70-90% between subunit isoforms. The largest areas of structural diversity are found within the large intracellular domains between TM3 and TM4.

The structural diversity of $GABA_A$ receptor subunits is further enhanced by alternative splicing of the α_6, β_2, β_4, and γ_2 subunits mRNAs in a number of species. In most instances, with the exception of the α_6 subunit, the structural diversity generated by these splicing events occurs within the predicted major intracellular domains of these subunits. For the γ_2 subunit, an insertion of eight amino acids within the TM3/TM4 cytoplasmic loop distinguishes two forms of the subunit, termed γ_{2S} and γ_{2L}. This insertion contains a serine residue that satisfies the consensus for phosphorylation by a number of protein kinases, including PKC. Alternative splicing of the β_2 subunit within the TM3-TM4 cytoplasmic loop is found in chicken and human mRNA but does not appear to occur in rodents.

Sizes of the alternatively spliced exons vary between these two species. In chickens an additional exon encoding 17 amino acids differentiates the β_{2S} and β_{2L} subunits, whereas in humans a much larger exon encoding 38 amino acids is found. These insertions occur at identical positions, and both insertions contain consensus sites for phosphorylation; most notably the insertion in the human β_2 subunit encodes a strong consensus for PKA phosphorylation.

Diversity of $GABA_A$ receptor structure in the brain has been examined by *in situ* hybridization and immunolocalization methodologies and the results suggest a large developmental and regional heterogeneity of $GABA_A$ receptor structure. In addition, individual neurons, for example, cerebellar granule neurons, have been demonstrated to express multiple receptor subunits, suggesting that some neurons may have the capacity to express multiple types of $GABA_A$ receptors. Expression of receptor cDNAs has been used to determine the minimal subunit requirement for the production of GABA-gated chloride channels that show the full pharmacological repertoire of neuronal $GABA_A$ receptors. Expression of α and β subunits supports GABA-gated currents that are modulated by barbiturates and inhibited by GABA antagonists and zinc ions but are not enhanced by benzodiazepines.

Coexpression of α, β, and either γ_2 or γ_3 subunits is crucial in conferring sensitivity to benzodiazepine and relative insensitivity to zinc antagonism on recombinant receptors. The α subunits appear to be important for determining benzodiazepine affinity. The α_1 subunit imparts type I benzodiazepine pharmacology, whereas the α_2 and α_3 subunits impart type II pharmacology, based on the differential affinity for the selective benzodiazepine CL 218817. The α_5 subunit imparts a novel type II pharmacology, whereas the α_6 subunit produces receptors with high affinity for the alcohol antagonist Ro15-4513 and low affinity for classical benzodiazepines such as diazepam. Expression of single subunits or the binary combinations $\alpha\gamma_2$, $\beta_1\gamma_2$ or $\beta_2\gamma_2$ has produced conflicting results. Recent cell biological studies have begun to clarify these electrophysiological observations.

The subcellular distribution of $GABA_A$ receptors constructed from α_1 β_2 and γ_{2L} subunits demonstrates that the dimer subunit combinations of $\alpha_1\gamma_{2L}$ and $\beta_2\gamma_{2L}$, although capable of oligomerization, are preferentially retained in the endoplasmic reticulum, as are the α_1, β_2, and γ_{2L} subunits when expressed alone. The only subunit combinations that appear to access the cell surface significantly are $\alpha_1\beta_2$ and $\alpha_1\beta_2\gamma_{2L}$.

Electrophysiological experiments from cells expressing the α_1, β_1, and γ_{2S} subunits have revealed that receptors incorporating all three subunits are preferentially expressed on the cell surface. Therefore the consensus of opinion suggests that *in vivo* $GABA_A$ receptors consist of α, β, and γ subunits.

Glycine receptors

Glycine is an established neurotransmitter in the spinal cord and brain stem where it mediates the inhibition of neuronal firing via GlyR activation. GlyRs, purified from rat spinal cord using strychnine affinity chromatography, are pentameric and consist of ligand-binding α (48 kDa) and structural β (58 kDa) subunits. In addition a receptor-associated protein of 93 kDa, termed gephyrin, copurifies with these subunits.

Cloning studies identifying the α_1 subunit of the glycine receptor have revealed that GlyRs share common structural features with both $GABA_A$ receptors and AChRs. The heterogeneity of GlyRs, inferred from the discovery of strychnineinsensitive GlyRs in neonatal spinal cord, has been confirmed by the isolation of additional cDNAs encoding GlyR subunits, designated α_1-α_4 and β with splice variants of some of these subunits. *In situ* hybridization and biochemical studies have shown that the GlyR α_1 subunit is abundant only postnatally and exists in combination with the GlyR β subunit; these two subunits together form the major GlyR isoform in adult spinal cord and brain stem.

Heterologous expression of α_1, α_2, or α_3 subunits alone generates chloride currents that can be gated by glycine, β-alanine, and taurine and inhibited by the antagonist picrotoxin. Although the β subunit cDNA alone is weak in directing the formation of GlyR channels, its coexpression with α subunits results in the production of heterooligomeric receptors that are picrotoxin-insensitive. Because the single channel properties of these receptors closely resembles those recorded from spinal cord neurons, native glycine receptors are generally considered to contain both α and β subunits.

Phosphorylation sites within $GABA_A$ receptors

Major intracellular domains of many $GABA_A$ receptor subunits contain a number of consensus sites for serine/ threonine and tyrosine protein kinases. β subunits contain a conserved site for a number of second messenger-dependent protein kinases, including PKA, corresponding to S409 in the case of β_1 subunit. The γ_{2L}. subunit encodes a site for phosphorylation that conforms to the consensus for a number of protein kinases, including PKC. Both the γ_1 and γ_2 subunits contain conserved consensus sites for tyrosine phosphorylation (Y365 and Y367 in the case of the murine γ_{2L} subunit). The α_6 subunit is unique among $GABA_A$ receptor α subunits in encoding a strong consensus site for phosphorylation by a number of kinases, including PKA.

Benzodiazepine affinity-purified preparations of $GABA_A$ receptors have been shown to be phosphorylated by a number of differing protein kinases. PKA and PKC both appear to phosphorylate "β subunits" based on apparent molecular masses (53-57 kDa) observed by sodium dodecyl sulphate polyacrylamide gel electrophoresis. Further evidence that these bands are indeed the receptor "β subunits" followed the observation that antisera directed against the intracellular domain of the β_1 subunit blocks the phosphorylation of these bands by both PKA and PKC. A presumed receptor-associated kinase, which is not stimulated by either phorbol esters or cyclic nucleotides, phosphorylates an "α sub unit". Recent studies have also demonstrated that purified receptors are substrates for the well-characterized tyrosine kinase v-src (the transforming gene product of the Rous sarcoma virus), which phosphorylates "β and γ subunits" as determined by migration on SDS-PAGE. Because of the heterogeneous nature of affinity-purified receptor preparations and the scarcity of $GABA_A$ receptors in the brain the identity of the subunits phosphorylated in these studies remains uncertain.

Recombinant expression of the intracellular domains of receptor subunits has been utilized to overcome the problems associated with identifying phosphorylation sites within neuronal $GABA_A$ receptor subunits. These regions are highly soluble when expressed as glutathione-S-transferase (GST) fusion proteins in *Escherichia coli*, allowing purification under native conditions. Using this approach it has been possible to identify high affinity kinase substrates in the intracellular domains of some $GABA_A$ receptor subunits. The murine β_1 subunit intracellular domain is phosphorylated to a high stoichiometry with high affinity by PKA, PKC, cGMP-dependent protein kinase (PKG), and calcium/calmodulin-dependent protein kinase II (CaM-kinase) on S409. The conserved residue (S410 in the case of the β_2 subunit) is also phosphorylated by these four kinases in both the β_2 and β_3 subunits. This site, R—R—R—(Aβ_1,β_2,Sβ_3)—S—L—Q—K (the major kinase substrate site is underlined), is found in all β subunits isolated from a variety of vertebrate species and in a β subunit isolated from the invertebrate *Lymnaea*.

There are additional sites in the murine β_1 and β_3 subunits for CaM-kinase II phosphorylation, identified as S383 and S384, respectively. The intracellular domain of the β_1 subunit can also be phosphorylated by *v-src in vitro*; however, the site(s) of phosphorylation have not been determined. These observations suggest a critical role for receptor β subunits in conferring cellular regulation upon $GABA_A$ receptors utilizing multiple cell signalling pathways.

Phosphorylation of both forms of the γ_2 subunits has also been analysed using similar methodologies. These studies reveal that S343, contained within the eight amino-acid insertion that differentiates these two forms of the γ_2 subunit, is a high-affinity substrate of both PKC and CaM-kinase II. In addition, both the γ_{2S} and the γ_{2L} intracellular domains are phosphorylated on additional residues by each of these kinases—PKC phosphorylates S327, whereas CaM-kinase II phosphorylates both S348 and T350 McDonald

Recent studies have also demonstrated that the γ_{2L} intracellular domains can be phosphorylated by v-src; however, the identity of the residues phosphoryiated in this study have not been reported. Overall these *in vitro* studies suggest that the β and γ_2 subunits are the major sites of phosphorylation within $GABA_A$ receptors *in vivo*. In parallel with these *in vitro* studies, the phosphorylation of $GABA_A$ receptor subunits has been investigated in heterologous expression systems such as the human embryonic kidney cell line (HEK).

After prelabeling expressing cells with [^{32}P]-orthophosphoric acid, receptor phosphorylation can be assessed by immunoprecipitation with subunit-specific antisera. Receptors can also be coexpressed with constitutively active protein kinases, such as the catalytic subunit of PKA, to produce constitutively phosphorylated receptor subunits. Murine $GABA_A$ receptors composed of either $\alpha_1\beta_1$ or $\alpha_1\beta_1\gamma_{2S}$ subunits expressed in HEK cells are phosphorylated specifically by PKA on S409 within the intracellular domain of the β_1 subunit. The β_2 and β_3 subunits are also phosphorylated on S410 and S409, respectively, by PKA when expressed with α_1 and γ_2 subunits in HEK cell.

Phosphorylation of receptors composed of $\alpha_1\beta_1$ $\alpha_1\beta_1\gamma_{2S}$, and $\alpha_1\beta_1\gamma_{2L}$ subunits by PKC has also been investigated. The a isoform of PKC specifically phosphorylates S409 in the β_1 subunit in these receptor subunit combinations. These studies of β subunit phosphorylation expressed in HEK cells are therefore in agreement with the *in vitro* phosphorylation studies performed on the cytoplasmic loops of receptor β subunits.

Studies of the phosphorylation of the γ_2 subunit have proven difficult. This subunit appears to be extremely susceptible to proteolysis; furthermore, it has been difficult to generate high affinity antisera against this subunit. These problems have recently been overcome by epitope tagging of the γ_2 subunit. The addition of reporter epitopes within the N-terminal domain of the γ_2 subunit has enabled the tyrosine phosphorylation of the subunit to be analysed when expressed in HEK cells. This cell line has very low steady-state levels of phosphotyrosine, indicating low intrinsic tyrosine kinase or high tyrosine phosphatase activity. This conveniently allows $GABA_A$ receptor tyrosine phosphorylation to be studied by coexpression of activated tyrosine kinases.

Coexpression of $GABA_A$ receptors consisting of $\alpha_1\beta_1$ and γ_{2L} subunits with v-src results in tyrosine phosphorylation of the γ_{2L} subunit on residues Y365 and Y367. The β_1 subunit is also phosphorylated, but to a much lower stoichiometry, on residues Y370 and Y372. In similarity with the sites for serine phosphorylation, the sites for tyrosine phosphorylation are found within the predicted intracellular domain of these two subunits.

$GABA_A$ receptors are the substrates of a number of different protein kinases, and there is significant interest in determining the functional effects of these phosphate modifications to receptor structure. The effects of PKA phosphorylation on $GABA_A$ receptor function have been examined in a number of different experimental systems, with complex and often contradictory results. Thus PKA appears to regulate $GABA_A$ receptor desensitization and to decrease receptor activation in cortical and spinal neurons and synaptoneurosomes but not in spinal cord neurons.

However, in the retinal and cerebellar Purkinje cells, vasoactive intestinal peptide (VIP) and noradrenaline enhance $GABA_A$ responses. These results suggest that PKA phosphorylation may enhance or inhibit $GABA_A$ receptor function in different neuronal types.The variability of the effects may reflect an underlying heterogeneity of receptor structure, suggesting a differential modulation of $GABA_A$ receptor function by PKA-mediated phosphorylation. This could, in turn, be dependent on receptor subunit composition.

However, these results are complicated by the observation that many hydrophobic drugs used to modulate protein kinase activity (such as forskolin) have direct effects on $GABA_A$ receptors that are independent of phosphorylation. Clarification of these issues requires the parallel measurement of receptor phosphorylation with receptor function. This, at present, is best achieved by the heterologous; expression of $GABA_A$ receptor cDNAs, which allows the precise effects of phosphorylation to be determined utilizing site-directed mutagenesis. Given the problems associated with membrane permeable second messenger modulators, a preferable strategy to modulate receptor phosphorylation in these studies is via intracellular dialysis with the catalytic subunit of PKA or with cAMP.

The functional effects of PKA-mediated phosphorylation have been studied on $GABA_A$ receptors, composed of either $\alpha_1\beta_1$ or $\alpha_1\beta_1\gamma_{2S}$ subunits, expressed in HEK cells. Phosphorylation of S409 in the β_1 subunit mediated by intracellular dialysis with either cAMP or the catalytic subunit of PKA results in a time-dependent decrease of GABA-induced currents and a modulation of $\alpha_1\beta_1$ receptor desensitization. The magnitude of inhibition is greater for receptors composed of $\alpha_1\beta_1\gamma_{2S}$ subunits, and these functional effects can be abolished by mutation of S409 to an alanine residue in the β_1 subunit, the sole site of PKA phosphorylation in these receptors.

Using the catalytic subunit of PKA it has been possible to demonstrate similar modulation of $GABA_A$ receptor function in cultured superior cervical ganglia (SCG), spinal cord neurons, and cerebellar granule cells. GABA-induced chloride flux from brain microsacs is also reduced by inclusion of the catalytic subunit of PKA. These findings with recombinant receptors do not explain the enhancements of $GABA_A$ receptor function seen in the retinal and cerebellar Purkinje cells, in which short-term exposure to VIP and noradrenaline, respectively, has been shown to enhance $GABA_A$ receptor currents.

Because the predominant effect of these agents is to increase intracellular levels of cAMP, this effect is likely to be mediated by PKA. Interestingly, the effects of VIP may be dependent on the species used because, in rabbit retina, VIP inhibits GABA-activated currents. In cerebellar Purkinje cells the effect of noradrenaline can be mimicked by cAMP analogues and blocked by the specific PKA inhibitor peptide (PKIP). This enhancement of $GABA_A$ receptor function, or "rebound potentiation," can also be induced by glutamate receptor activation and elevated levels of intracellular Ca^{2+}.

Likewise, in rat retina, intracellular dialysis with the catalytic subunit of PKA enhances $GABA_A$ receptor responses, but agents that activate PKA do not affect $GABA_A$ receptor function in retinal cells prepared from rabbits. Recent studies have shown that dopamine, but not noradrenaline, is capable of enhancing $GABA_A$

receptor function in the retina. This observation is of interest because both these neurotransmitters are capable of activating PKA. The reasons for the apparently differing effects of PKA phosphorylation in these neurons compared to SCG, spinal cord, cerebellar granule neurons, and some recombinant preparations will require further experimentation.

Differences in the species of animals used, culture conditions, and recording methodologies may be of importance. Receptor heterogeneity may underlie these differing modes of regulation, and this problem can be most conveniently addressed using recombinant methodologies. However, direct receptor phosphorylation by PKA may not be responsible for some of these effects, which may result from the activation of other cell signalling cascades that cause a differential modulation of receptor function.

In addition to effects on channel function, chronic activation of PKA has been reported to enhance the assembly of $GABA_A$ receptors. This finding comes from the expression of $GABA_A$ receptors in cell lines that have various levels of constitutive (cAMP-independent) PKA activity. Angelotti *et al.* used three cell lines: Cα12, L929, and RAB10, which have PKA activities of 500, 100, and 5 units/mg protein, respectively. Larger GABA responses were seen from receptors composed of $\alpha_1\beta_1$ and γ_{2S} subunits in Cα12 cells compared to the other two cell lines but not from receptors composed of only α_1 and β_1 subunits. This effect could be blocked by mutation of S409 in the β_1 subunit, the sole site of phosphorylation in receptors composed of $\alpha_1\beta_1$ and γ_{2S} subunits.

Experiments with expressed $GABA_A$ receptors produced after injection of either rat or chick brain mRNA into *Xenopus* oocytes provided the first evidence that $GABA_A$ receptors were subject to modulation by

PKC-mediated phosphorylation. In these pioneering experiments, activation of PKC using phorbol esters resulted in a significant inhibition of GABA-induced whole-cell currents. This observation has been further examined by utilizing heterologous expression of receptor cDNAs. These initial studies have shown that phorbol ester- induced PKC activity can inhibit the function of a range of receptors constructed from α1-α5, β1-2, and γ_2 subunits. In the case of receptors composed of $\alpha_1\beta_1\gamma_2$, subunits, the effects of phorbol esters can be blocked by PKC inhibitory, peptide (PKCI). The role of specific phosphorylation sites for PKC within the predicted intracellular domains of these subunits, as defined by biochemical methodologies, has been examined by site-specific mutagenesis.

These studies have concentrated on the significance of the additional phosphorylation site for PKC within the eight amino-acid insertion that differentiates the two forms of the γ_2 subunit. The phosphorylation site is of special significance because ethanol is believed to exert much of its action on $GABA_A$ receptor function via this residue. PKC exerts its inhibitory effect on receptors composed of $\alpha_1\beta_1$ or β_2 Subunits, either with or without the γ_{2L} or γ_{2S} subunits. The effects occur via multiple phosphorylation of S409 within the β_1 subunit, S327 in both the γ_{2L}/γ_{2S} subunits, and S343 within the γ_{2L} subunit.

The effects of phosphorylation are also related to the GABA concentration, with larger inhibitions apparent with receptors incorporating the γ_{2L} subunit at high concentrations of GABA. Selective mutagenesis reveals that phosphorylation at any of the sites on the β_1 or γ_2 subunits is sufficient to produce negative modulation. However, phosphorylation at S343, which is contained within the eight extra amino acids within the γ_{2L} subunit, produces the largest inhibitory effect, suggesting that the phosphorylation sites are not functionally equivalent. Moreover, there is no apparent linear relationship between the degree of inhibition and the number of phosphorylation sites. $GABA_A$ receptors composed of α_1 and β_1 subunits are also subject to negative modulation by PKC-mediated phosphorylation.

PKC phosphorylation also modulates rapid desensitization of receptors composed of these subunits, as previously described for PKA-mediated phosphorylation. This is not unexpected be cause S409 represents the sole site of phosphorylation for both PKA and PKC in these receptors. Phorbol ester treatment has been used to examine the regulation of neuronal $GABA_A$ receptors by PKC phosphorylation. Responses in SCG neurons are inhibited by phorbol ester treatment but not by α-phorbols, which are incapable of activating PKC. GABA-induced chloride flux from cerebellar microsacs can be selectively inhibited by PKC activators, but PKC does not appear to modulate receptor desensitization in spinal cord microsacs.

Recent experiments have shown that synaptic currents mediated via $GABA_A$ receptors can be regulated by a PKC-dependent mechanism in CA1 hippocampal neurons. Inclusion of the specific peptide PKC inhibitor PKCI enhances $GABA_A$-mediated inhibitory postsynaptic potentials. This is in agreement with experiments demonstrating the inhibition of receptor function by direct PKC-mediated phosphorylation.

Recent experiments have also shown that bicuculline-insensitive receptors expressed in rat retinal bipolar cells can be inhibited by a PKC-dependent process. In addition to using phorbol esters to modulate receptor phosphorylation, intracellular dialysis of trypsin-cleaved PKC from rat brain has also been used. Cleavage with trypsin leads to the constitutive activation of PKC. Intracellular application of this preparation at a concentration of 40 nM to L929 cells transiently expressing $GABA_A$ receptors composed of α_1, β_1, and γ_{2L} subunits enhances receptor function. This enhancement can be blocked by the PKCI peptide. Whether these effects are due to direct receptor phosphorylation by PKC, however, has not been examined. This information is important because purified preparations of rat brain PKC contain a broad range of PKC isoforms, which play central roles in controlling a variety of signal transduction pathways.

Functional effects of these two kinases on $GABA_A$ receptor function is poorly understood. $GABA_A$ receptor-mediated currents in neurons from the nucleus of the tractus solitarius of the rat are inhibited by cGMP or on the activation of metabotropic glutamate receptors. Ca^{2+} inhibits or activates GABA-mediated currents in different neuronal populations, but the role of CaM-kinase II in these actions of Ca^{2+} is not well characterized. However, recent experiments have shown that constitutively activated CaM-kinase II can activate $GABA_A$ receptor-mediated whole-cell currents in cultured rat spinal cord neurons. Whether these effects are mediated by direct receptor phosphorylation by CaM-kinase II has not been determined.

Role of tyrosine phosphorylation mediated by v-src on the functional properties of $GABA_A$ receptors composed of $\alpha_1\beta_1$ and γ_{2L} subunits expressed in HEK cells has recently been reported. Tyrosine phosphorylation of Y365 and Y367 in the γ_{2L} subunit, identified as the major src substrates in $GABA_A$ receptors composed of these subunits, enhances GABA-induced currents. In agreement with this, Valenzuela *et al.*, have shown that tyrosine kinase inhibitors reduced the magnitude of GABA-gated currents recorded from *Xenopus* oocytes expressing

$GABA_A$ receptors composed of $\alpha_1\beta_1\gamma_{2L}$ and $\alpha_1\beta_1$ subunits. Neuronal $GABA_A$ receptors can also be modulated by tyrosine phosphorylation. Whole-cell currents recorded from cultured SCG neurons are reduced by exposure to tyrosine kinase inhibitors and are enhanced by tyrosine phosphatase inhibitors.

Furthermore, GABA-mediated chloride flux from brain microsacs is also reduced by tyrosine kinase inhibitors. Single-channel recording from SCG neurons suggests that tyrosine phosphorylation increases both the mean open time and also the probability of channel opening. These results indicate that tyrosine phosphorylation may be a means of enhancing or maintaining $GABA_A$ receptor function. Interestingly, phosphorylation by unidentified kinases (neither PKA nor PKC), or "phosphorylation factors," has been implicated in preventing the washout or rundown of $GABA_A$ receptor responses in a number of neurons.

Electrophysiological studies have been used to determine if defined neurotransmitters can modulate the functional properties of $GABA_A$ receptors. Interestingly, the activation of serotonin and glutamate receptors can modulate $GABA_A$ receptor function in a number of neuronal systems. Activation of 5-HT_7 receptors in rat suprachiasmatic neurons resulted in decreased $GABA_A$ receptor function via a cAMP-dependent process. The effects of glutamate receptor activation on $GABA_A$ receptor function have been analysed in hippocampal neurons. Early observations suggested that the activation of both NMDA and non-NMDA types enhanced $GABA_A$ receptor currents recorded from the CA1 region.

Recent findings have refuted this initial observation and, in contrast, have demonstrated that activation of NMDA receptors leads to a Ca^{2+}- and PKC-dependent decrease in $GABA_A$ receptor function. The identity of the signalling pathways involved is unclear, but biochemical measurements of the state of $GABA_A$ receptor phosphorylation should help clarify these observations. There is general agreement that $GABA_A$ receptors are the substrates for multiple protein kinases. The β and γ_2 subunits appear to be the major substrates for phosphorylation by PKA, PKC, PKG, CaM-kinase II, and the prototypic tyrosine kinase, src. The functional effects of receptor phosphorylation, however, appear complex.

Recombinant studies have demonstrated that direct phosphorylation by either PKA or PKC inhibits receptor function, whereas PKA phosphorylation may also be important in receptor assembly. A similar regulation of neuronal $GABA_A$ receptor function by PKA and PKC phosphorylation has been reported. In some neuronal preparations, however, PKA-mediated phosphorylation increases receptor function.

These discrepancies may be due to differences in receptor structure or the activation of other signalling pathways. The functional effects of PKC induced phosphorylation on both neuronal and recombinant receptors are almost exclusively inhibitory when phorbol ester treatment is used to activate PKC. In contrast, trypsin-cleaved PKC from rat brain appears to enhance GABA-induced currents when receptors are composed of α_1, β_1, and γ_{2L} subunits expressed in L929 cells. In comparison, tyrosine phosphorylation of sites on the γ_2 subunit appear to enhance receptor function of recombinant receptors composed of $\alpha_1\beta_1$ and γ_{2L} subunits. Similar modulation of receptor function is seen in cultured SCG neurones. These results suggest that this process may be important in enhancing or maintaining receptor function.

Recent observations with tyrosine kinases are also important because of the varying reported effects of PKC on receptor function. Recent studies have identified a novel protein tyrosine kinase, homologous to focal adhesion kinase, called PYK2 or CAKβ that is rapidly activated both by elevations in intracellular calcium and by PKC. Northern blotting has revealed that this kinase is widely expressed in the brain and in a number of cell lines, and it has been shown to phosphorylate and regulate inwardly rectifying potassium channels.

Because tyrosine phosphorylation by v-src enhances receptor function, some of die variable effects seen on $GABA_A$ receptor function following the activation of PKC may be mediated by PYK2. Interestingly preliminary studies from our laboratories demonstrate that the γ_2 subunit is phosphorylated on tyrosine residues by this kinase. To test the relevance of these regulatory mechanisms *in vivo* requires the overexpression of mutated receptor subunits devoid of phosphorylation sites for selected protein kinases in neuronal networks. Such studies can be performed with transgenic and viral methodologies and should help clarify our understanding of $GABA_A$ receptor modulation.

The role of phosphorylation in modulating the functional properties of GlyRs is less well characterized. Strychnine affinity- purified preparations of GlyRs can be phosphorylated to high stoichiometry by purified PKC *in vitro* within the 42-kDa strychnine-binding subunit. The major site of phosphorylation by PKC in these purified preparations has been identified as S390 within the α_1 subunit. This residue is found within the predicted major intracellular domain, in common with phosphorylation sites located within both $GABA_A$ and muscle nicotinic AChR subunits as described previously.

Interestingly, GlyR ligands can modulate the rate of receptor phosphorylation by PKC. GlyRs can also be phosphorylated in spinal cord neurons following the activation of both PKC and PKA. The site (s) of phosphorylation for PKA, however, have not been elucidated. The functional effects of GlyR phosphorylation have been examined in a number of experimental systems.

Glycine responses in cultured spinal neurons are enhanced by activators of PKA. In contrast to this, glycine responses recorded from both dissociated neurons derived from either the substantia nigra or the ventromedial neurons are inhibited by activators of PKA. GlyRs expressed in *Xenopus* oocytes following injection with mRNA extracted from spinal cord can be differentially regulated by PKA and PKC; PKA activation appears to enhance receptor function, whereas PKC activation is inhibitory. However, recent data have also suggested that GlyR function can be enhanced by PKC activation in hippocampal neurons. This effect of PKC activation can be mimicked by activation of 5-HT receptors.

Heterologous expression of GlyR cDNAs has provided the strongest experimental evidence that GlyRs can be modulated by direct phosphorylation. Glycine responses from homomeric α_1 GlyRs are inhibited by activation of PKC, and this regulation is abolished by the mutation of S390, which has been previously identified as a substrate of PKC in affinity-purified preparations of GlyRs. It thus appears from these studies that the functional effects of phosphorylation on GlyRs are diverse. PKA activation is capable of enhancing, as well as inhibiting, the functional properties of these receptors in differing neuronal preparations.

As with $GABA_A$ receptors, these differential functional effects of PKA phosphorylation on GlyRs may be due either to receptor heterogeneity or to the activation of other signalling cascades.

Excitatory amino acid receptors

Excitatory amino acid (EAA) receptors are referred to as glutamate receptors although glutamate may not be the only natural transmitter. EAA receptors were originally classified according to their sensitivity to the three agonists kainic acid (KA), α-amino-3hydroxy-5-methyl-4-isoxazolepropionic acid (AMPA), and N-methyl-D-aspartate (NMDA), resulting in three populations of excitatory receptors that are also sensitive to glutamate. These receptors are ligand-gated ion channels (ionotropic) and are referred to as NMDA and non-NMDA receptors. Recently, an additional class of excitatory receptors was revealed that is coupled to G proteins and classified as the metabotropic receptor.

The application of molecular studies to EAA receptors has revealed a level of heterogeneity far greater than that predicted from pharmacological studies. This has resulted in a new classification for those protein subunits that form functional excitatory amino acid receptors. Non-NMDA receptors exhibiting a higher affinity for AMPA than for KA are designated GluRl-GluR4; receptors displaying a low affinity for KA, which, nevertheless, is higher compared to that for AMPA are designated GluR5-GluR7; and receptors with a much higher affinity for KA than for AMPA are classified as KA-1 and KA-2.

Many, but not all of these polypeptides can be expressed in homomeric or heteromeric combinations. The GluR1-GluR4 subunits can all exist in one of two splice variant forms defined as *flip* or *flop*. In addition, GluR4 can also exist as a GluR4C spliced variant. The NMDA receptor family has one key subunit, NMDARI (alternative nomenclature, ζ1), which can be expressed as a homomeric receptor or in combination with other subunits selected from NMDAR2A-NMDAR2D (alternatives, NR2A-NR2D or ε1-ε4). The series NMDAR2A-NMDAR2D cannot form homomeric receptors or heteromeric receptors with other NMDAR2 members.

The NMDAR1 subunit can exist in a variety of splice variants to further increase heterogeneity. Metabotropic glutamate receptors consist of a family of eight discrete subtypes defined as mGluRl-mGluR8. Splice variants have been reported for three metabotropic glutamate receptors: mGluR1, mGluR4, and mGluR5.

According to the transduction mechanisms regulated by the mGluRs, these receptors can be classified into three groups: mGluR1 and mGluR5 initiate phosphoinositide hydrolysis; mGluR2 and mGluR3, are negatively coupled to adenylate cyclase; and mGluR4 and mGluR6-mGluR8 are positively coupled to adenylate cyclase.

AMPA/KA receptors

One of the first reports implicating the involvement of phosphorylation in the modulation of glutamate receptor function came from studies of white perch (*Roccus americana*) retina. Responses activated by KA or glutamate were enhanced by dopamine. Dopamine had little direct effect on the resting membrane conductance but activated adenylate cyclase leading to increased levels of cAMP. The mediation by cAMP of the effect of dopamine on KA-activated responses was supported by the membrane permeant derivative 8-bromo-cAMP, producing similar effects in the absence of dopamine. In white perch cultured retinal horizontal cells, single channel recording and membrane current fluctuation analysis of EAA gated ion channels revealed that dopamine increased the frequency of 5-10 pS channel events and increased the probability of channel opening without affecting the number of functional channels or the single channel current amplitudes. Dopamine also reduced receptor desensitization to glutamate, quisqualate, or AMPA, an effect occluded by prior removal of desensitization by aniracetam. Interestingly, kainate-activated currents were relatively unaffected. At the level of the synapse, the glutamate receptor-mediated excitatory postsynaptic current (EPSC) recorded from goldfish Mauthner cells was also enhanced by dopamine, an effect blocked by SCH-23390 or by postsynaptic intracellular injections of the cAMP- dependent protein kinase inhibitor PKI_{5-24}, indicating a role for protein phosphorylation.

In the mammalian CNS, whole-cell currents evoked by glutamate or kainate on cultured hippocampal pyramidal neurons were enhanced by forskolin, an activator of adenylate cyclase, but not by the inactive analogue 1,9-dideoxyforskolin. The involvement of PKA was suggested by the ability of PKI_{20}amide to apparently inhibit the effect of forskolin.

Moreover, PKI_{20}amide alone reduced the response amplitude to glutamate, which is compatible with some basal level of PKA activity. Single-channel analysis, using outside-out patches exposed to glutamate and including the catalytic subunit of PKA in the patch pipette solution with ATP, resulted in an increased frequency of channel opening and mean open time for non-NMDA channels but had no effect on NMDA-activated channels. Synaptic glutamate receptors were analysed by recording miniature EPSCs. Forskolin increased the amplitude and decay time for the miniature EPSCs but not in neurons dialysed with PKI_{20}amide. Similar results, in which the catalytic subunit of PKA prevented the rundown in the amplitude of the KA-activated response, were reported by Wang and colleagues.

Furthermore, incorporation of cAMP in the recording pipette enhanced the response to KA but not to

NMDA. Endogenous PKA was thought to modulate KA-activated currents after using the nystatin perforated-patch technique in conjunction with membrane permeant cAMP analogues. Sp-cAMPS (an activator of PKA) potentiated the KA response whereas Rp-cAMPS (a competitive inhibitor of PKA) prevented the enhancement caused by Sp-cAMPS and reduced the basal current amplitudes evoked by KA. The predicted balance between phosphorylation and dephosphorylation was analysed using the protein phosphatase 1 and protein phosphatase 2A inhibitor okadaic acid. This agent resulted in enhanced responses to KA and, because okadaic acid has little effect on acid and alkaline phosphatases and phosphatase 2C, these enzymes are probably not involved in the response modulation.

Long-term potentiation (LTP) has featured prominently as a potential molecular mechanism for learning and memory. It is reliant on intact excitatory synapses undergoing an increase in their efficacy after nerve fiber stimulation, which can be long lasting and may involve Ca^{2+}- and phospholipid-dependent protein kinase (PKC). Although the precise mechanistic details have yet to be ascertained, both NMDA and non-NMDA receptors can play a pivotal role in the generation and maintenance of LTP. One conceivable way in which excitatory "synaptic efficacy" could be enhanced over long periods is by phosphorylation-dependent mechanisms enhancing the responses transduced by glutamate receptors.

Thus, non-NMDA receptors were considered a likely substrate for PKC. Kainate responses recorded from hippocampal neurons were enhanced by internal dialysis with the constitutively active catalytic fragment of PKC, designated PKCM, which becomes insensitive to Ca^{2+} and phosphatidylserine. This effect was blocked by the PKC inhibitory peptide PKCI (19-36), which also slightly inhibited the responses to KA, suggesting basal PKC-induced phosphorylation of non-NMDA receptors.

The effect of PKCM was dependent on the concentration of KA, with responses induced by concentrations higher than the KA EC_{50} (approximately 30 μM) being enhanced by PKCM, whereas responses evoked by low concentrations of KA (less than 30 μM) were inhibited. Similarly, okadaic acid produced an inhibition at low KA concentration and an enhancement of KA responses at high KA concentrations.

After internal dialysis with PKCM, non-NMDA receptor-mediated miniature EPSCs, evoked using periodic release of a hypertonic solution near the recorded neuron, exhibited either prolonged decays or increased current amplitudes. These effects occurred concurrently with an increased postsynaptic sensitivity to KA.

The complex differential modulation by PKCM of hippocampal non-NMDA receptors, revealed from the dose-response curve analysis, has many interpretations. These include the presence of distinct non-NMDA receptor subtypes with differential sensitivities to PKCM, an alteration by PKCM in the apparent affinity of the receptor, or a change in the cooperativity of receptor activation. Experiments using hippocampal brain slices suggested that activation of PKC using phorbol esters resulted in enhanced excitatory synaptic transmission resulting from increased contributions of both the non-NMDA and NMDA components. The NMDA receptor-mediated EPSC appeared to be most affected by PKC; the enhancement was inhibited by extracellular application of the PKC inhibitor H-7 but not by postsynaptic dialysis with H-7. This latter observation is consistent with phorbol esters acting via a presynaptic site and increasing the release of the excitatory transmitter, although the postsynaptic sensitivities to exogenously applied NMDA and non-NMDA agonists were not assessed. In contrast to the effects of PKC on hippocampal non-NMDA receptors, both KA- and AMPA-activated currents in trigeminal nucleus neurons were unaffected by intracellular injection of PKC.

Protein kinase C activation has been linked to the phenomenon of long-term synaptic depression (LTD) at parallel fiber-Purkinje cell synapses in the cerebellum. Crepel and Krupa noted that indirect measurements of glutamate-induced responses (extracellularly recorded spike firing frequency) in Purkinje neurons were reduced by application of phorbol esters.

Application of the inactive 4α-phorbol was without effect. Although the nature of the glutamate receptors involved in this study were not ascertained, the responses induced by aspartate were unaffected by phorbol esters. A patch clamp study of LTD in cerebellar neurons established that phorbol esters reduced the response to AMPA but did not affect NMDA-activated responses. The depression of the AMPA-induced current was prevented by calphostin C, a membrane permeable PKC inhibitor.

Early studies using mRNA injection and the expression of presumably heterogeneous populations of glutamate receptors have reported mixed results with regard to the modulatory effects of PKC. Using chick forebrain mRNA, Sigel and Baur found no effect of phorbol esters on KA-activated responses, whereas Moran and Dascal used rat brain mRNA and reported either no effect or an inhibition of AMPA-and KA-activated responses, depending on the concentration of 4β-phorbol 12,13-dibutyrate (0.1-0.5 μM). The synthetic diacylglycerol, 1-oleoyl-2-acetylglycerol (OAG), was without effect.

There are relatively few studies demonstrating that glutamate receptors can be phosphorylated by PKC. Phosphorylation of hippocampal glutamate receptors was assessed by exposure to [^{32}P]orthophosphoric acid and activation of PKC using phorbol 12-myristate 13-acetate [PMA, equivalent to 12-O-tetradecanoylphorbol 13-acetate (TPA)].

Native glutamate receptors were immunoprecipitated using antisera to GluR1 non-NMDA Subunits and subjected to phosphoamino acid analysis. Activation of PKC increased the level of phosphorylation that was associated with serine (73%) rather than threonine residues (27%). The specificity of PMA was not assessed, although phosphorylation was unaffected by the CaM-kinase inhibitor KN-62, and PMA also resulted in phosphorylation of the myristolated alanine-rich C kinase substrate MARCKS. Whether the PKC-induced phosphorylation occurred exclusively on the GluR1 subunit within heteromeric glutamate receptors or on other members of the AMPA receptor family, for example, GluR2-GluR4, remains to be established.

Phosphorylation of forebrain synaptosomal and postsynaptic density glutamate receptors (GluR1) by purified PKC has also been demonstrated. Immunoprecipitation studies have also indicated that GluR1 subunits in cortical neurons are phosphorylated under basal conditions on serine residues within a single phosphopeptide. The same phosphopeptide was phosphorylated under basal conditions in recombinant GluR1 subunits expressed in HEK cells. Interestingly, the overall level of phosphorylation of GluR1 in cortical neurons can be increased by phorbol esters and by spontaneous synaptic activity.

Immunoprecipitation of glutamate receptors using GluR1 antisera was also employed to study the degree and level of phosphorylation induced by CaM-kinase II CaM-kinase was activated either by applying glutamate and glycine to activate NMDA receptors, thereby permitting Ca^{2+} influx, or by using the divalent cation ionophore ionomycin.

Hippocampal neurons, forebrain synaptosomes, or postsynaptic densities exposed to ^{32}P revealed that phosphorylation occurred mostly on serine residues (68%) over threonine residues (32%), with no detectable phosphorylation of tyrosine residues. The cell permeable CaM-kinase inhibitor KN-62 blocked the phosphorylation induced after glutamate or glycine treatment suggesting some degree of specific activation.

The type of CaM-kinase involved was believed to be CaM-kinase II, mainly on the basis of the abundance of this isoform in the brain and its localization to the postsynaptic densities. Whether one or more GluR subunit(s) were phosphorylated by CaM-kinase was unresolved.

Responses activated by AMPA or kainate via non-NMDA receptors present on acutely isolated dorsal horn neurons were enhanced by internal dialysis with an α subunit of CaM-kinase II. Preactivation by autothiophosphorylation rendered the kinase approximately 70% Ca^{2+}-independent. Heat-inactivated kinase did not affect the non-NMDA agonist-induced responses, whereas dialysis with the nonselective phosphatase inhibitor microcystin enhanced agonist-induced responses.

CaM-kinase II was also introduced via an intracellular electrode into substantia gelatinosa cells in spinal cord slices, resulting in enhanced stimulus- evoked excitatory postsynaptic potentials (EPSPs) apparently independent of the neuronal input resistance. Kainate-activated currents in cultured hippocampal neurons were also enhanced by microcystin or by intracellular inclusion of preautothiophosphorylated CaM-kinase II. Nonphosphorylated or heat-inactivated CaM kinase II failed to affect the response to kainate. It was concluded that because GluR and CaM-kinase II colocalize at the postsynaptic densities, CaM-kinase is probably directly phosphorylating the GluR, although this has yet to be demonstrated. A role for CaM-kinase II in modulating synaptic AMPA receptors was suggested from the transient enhancement of EPSCs following calcium current activation in hippocampal CA1 pyramidal neurons. This enhancement was abolished by KN-62, which has no effect on the calcium current.

To assess the involvement of protein kinases, cells were bathed in the phosphatase inhibitor calyculin A, which resulted in a sustained potentiation of the EPSC following activation of the calcium current. Similar results were obtained by including okadaic acid or microcystin LR in the patch pipette, thereby only inhibiting phosphatases in the postsynaptic cell. This suggested that the level and duration of phosphorylation was linked to the enhanced EPSC amplitudes, presumably via phosphorylation of the receptor protein. This could be important in the long-term regulation of synaptic efficacy. In the absence of voltage pulses to activate calcium currents, the application of phosphatase inhibitors did not affect the EPSCs.

Previous studies on the potential of basal phosphorylation to regulate non-NMDA receptors in the presence of phosphatase inhibitors either showed some, enhancement or virtually no effect. One interesting finding by Wylie and Nicoll concerns the increased frequency of mEPSCs observed following the activation of calcium currents in the presence of a phosphatase inhibitor. The mechanism behind this increase is unclear but could reflect detection of hitherto unresolved events by virtue of the enhanced mEPSC amplitude.The enhancement of non-NMDA receptor-mediated responses by CaM-kinase II could form part of a synaptic positive-feedback loop. Essentially, calcium reflux could occur via glutamate-operated ion channels and also via calcium channels following the glutamate-induced cell depolarization. The influx of calcium would activate CaM-kinase II causing phosphorylation either of the glutamate receptors directly or of intermediary proteins, leading to enhanced receptor function protein phosphatases would then limit the duration and intensity of this effect.

Recombinant non-NMDA receptors

Modulation of non-NMDA receptor function either by pharmacological agents regulating the activity of protein kinases or by activated kinases and phosphatases alone, reinforced the concept that glutamate receptors are substrates for direct phosphorylation. Phosphorylation consensus sequences for protein kinases have been postulated for PKA, PKC, and CaM-kinase in different glutamate receptor subunits. These sites are located between die putative transmembrane domains, TM3 and TM4, which are postulated in some, but not all, models of membrane topology to be intracellular components of non-NMDA receptors. GluR1-GluR4 do not possess ideal phosphorylation consensus sequences for PKA; however, phosphorylation sites for PKC (serine 650 and threonines 716 and 732) and CaM-kinase (serine 627) have been observed. Furthermore, in GluR6, two serines at positions 684 and 666 have been identified as sites for PKA phosphorylation.

The consensus sequence for phosphorylation by PKA, identified in GluR6 non-NMDA receptors, was directly studied by transiently expressing homomeric

GluR6 subunits in HEK cells. Cotransfection with DNA encoding for the catalytic subunit of PKA (Cα), resulted in phosphorylation of GluR6 in the presence of ^{32}P. Increased phosphorylation was also apparent in cells exposed to forskolin.

The functional effect or phosphorylation (induced by internal dialysis with Cα or cAMP via the patch pipette) was to increase the response amplitude to glutamate. This effect was prevented by PKI_{5-24} amide. The location of the site of phosphorylation was suggested by the strong consensus sequence: R—R—X—S—V (where X represents a recognition-neutral amino acid) between presumed domains TM3 and TM4. Site directed mutagenesis of serine 684 to alanine (S684A) resulted in the abolition of phosphorylation and its functional consequences.

Although serine 684 represents the major site for PKA-induced phosphorylation, there may be additional minor sites that remain to be identified. Similar results were reported by Wang and colleagues for KA-activated responses on concanavalin A-treated GluR6 subunits. Site-directed mutagenesis of S684 did not completely abolish the functional effect of PKA. An additional mutation was made at S666 (consensus, F—K—K—S—K) and, although the double mutant GluR6 receptor was now insensitive to PKA, GluR6 receptors incorporating the single mutant, S666A, were still modulated by PKA. It is conceivable that S666 plays a structural role in the response modulation, which is dependent on phosphorylation at S684.

Overall, the major site of phosphorylation appeared to be S684. The mechanism of action for PKA has not yet been determined but it does not involve changes to the EC_{50} or the current-voltage relationship for the response to KA. Only responses to concentrations in excess of 10 μM KA seem to be enhanced by PKA. This is compatible with increased numbers of receptors or increased frequency of channel opening. Apart from GluR6 subunits, KA activation of GluR1 and GluR3 subunits expressed in *Xenopus* oocytes has also been reported to be enhanced by agents that increase the activity of PKA, although whether this involves phosphorylation is yet to be clarified.

The expression of GluR3 or GluR6 homomers or GluR1/GluR3 or GluR2/GluR3 heteromers in Xenopus oocytes resulted in KA-activated currents that were inhibited by PMA, probably by activation of protein kinase C. PMA also decreased the rate of decline in the response to KA, which the authors interpreted as a slow form of desensitization; however, part of this reduction could simply be due to the smaller KA response amplitude inducing less desensitization. Intracellular injection of PKCI into oocytes expressing GluR3 homomers prevented the PMA-induced decrease in the KA activated response, suggesting the phorbol ester was mediating its effect by activating PKC.

The control of receptor function by PKC is clearly complex because this expression study differs from the studies conducted on hippocampal neurons. As regards the recombinant receptor studies, a worthwhile control experiment to implicate the involvement of protein kinases and phosphorylation would be to mutate the residues that are likely to be phosphorylated. If this action removed the functional consequences of phosphorylation, it would constitute additional, but not absolute, evidence of a role for direct phosphorylation; however, site-directed mutagenesis does not discount the involvement of any intermediary protein(s) influencing receptor function. Following expression in Sf9 insect cells, GluR1 subunits can be phosphorylated by PKC with predominantly serine residues forming the target.

GluR1 subunits expressed in HEK cells exhibit some basal phosphorylation on serine residues; however, it is of interest that in one study activation of PKA or PKC did not consistently change the level of phosphorylation, suggesting that GluR1 might be a substrate for an unidentified serine or threonine kinase.

Homomeric GluR1 receptors, expressed in Sf9 cells, were also phosphorylated by CaM-kinase, largely on serine residues (87%) as determined by phosphopeptide mapping. Functional studies of expressed GluR6 receptors in HEK cells and GluR1 in *Xenopus* oocytes revealed that internal application of activated CaM-kinase produced enhancements of approximately 50% to KA-activated currents. The EC_{50} for KA on GluR1 receptors was unaltered. The site of phosphorylation was determined in GluR1 using site-directed mutagenesis.

Serine 627 fulfilled the consensus criteria for a CaM-kinase site containing an arginine three residues N-terminal to a serine or threonine, —R—M—V—S—. Mutation of serine 627 in GluR1 to alanine prevented the modulation by CaM-kinase, indicating that this is the likely residue subjected to phosphorylation.

The apparent affinity for KA and the inward rectification demonstrated by GluR1 receptors were unaltered by this mutation. A synthetic peptide, duplicating residues 620-638 and incorporating S627 of GluR1, was also phosphorylated by CaM-kinase on S627. This residue was unaffected by PKA or PKC.

The phosphopeptide map of the synthetic peptide appeared very similar to the phosphopeptide map obtained with hippocampal neuronal glutamate receptors after exposure to CaM-kinase. This suggests that GluR1 and S627 may feature prominently in the modulation of neuronal non-NMDA receptors by CaM-kinase.

NMDA receptors

Whole-cell or single-channel current activated by aspartate or glutamate were unaltered by cAMP or PKA suggesting that NMDA receptors, in contrast to non-NMDA receptors, in contrast to non-NMDA receptors, are not substrates for PKA. In comparison, NMDA receptors are thought to be substrates for PKC, but the actions of PKC activation have proved controversial. Whereas Linden and Connor reported that phorbol 12,13 diacetate (PDA) did not reduce the response to NMDA on cerebellar Purkinje neurons, enhancement of NMDA-activated responses has been reported following PKC activation in hippocampal, spinal cord, and trigeminal neurons. Activation of PKC enhanced the responses to NMDA in trigeminal neurons by increasing the probability of channel opening and also by reducing the voltage-dependent block by external, but not internal Mg^{2+}. The latter effect occurred by a PKC-induced increase in the apparent dissociation constant for Mg^{2+}. This action results in NMDA receptor-ion channels being activated at more negative membrane potentials and allows the prospect of increased Ca^{2+} influx via the NMDA ion channel, leading to increased PKC and eventually NMDA receptor-ion channel activity.

Conceivably, the modulation of NMDA receptors by PKC may vary in different brain areas. In cerebellar granule neurons PKC activation caused a reduction in the rise in intracellular calcium following NMDA receptor stimulation. A similar effect was observed with okadaic acid. Inactive 4a-phorbol ester or PMA in the presence of the inhibitors staurosporine or calphostin C led to little change in the NMDA response. The principal effect of PMA was to increase the EC_{50} for the coagonist glycine at the NMDA receptor. Increasing the glycine concentration overcame this effect. However, the proportion of Ca^{2+} influx via the NMDA ion channel compared to influx via voltage-gated calcium channels, which would be activated once the cells depolarized after exposure to NMDA, remains to be established. It is conceivable that phorbol esters may also have effects on calcium channel function that could complicate interpretation of the results.

An alternative explanation for the differential modulation of NMDA receptors by PKC in different brain areas could involve a dependence on the receptor subunit composition. With regard to synaptic glutamate receptors, the amplitude and duration of the EPSC in hippocampal CA1 neurons was increased by PDA. This was mostly due to an increase in the NMDA receptor-mediated component[6-cyano-7-nitroquinoxaline-2,3-dione (CNQX)-insensitive] of the EPSC. The effect was blocked by externally applied H-7 but not by internal perfusion with the same inhibitor. This evidence is in accordance with a presynaptic action of the phorbol esters causing an increase in transmitter release.

Use of H-7 as an inhibitor of PKC may pose some problems because one report suggested that it could interact *directly* with the NMDA receptor-ion channel. Interestingly, this action was observed only at the single channel level, in which H-7 reduced the number of bursts of channel activity and caused an increase in the number of brief closures. Overall, the burst length was increased such that the macroscopic NMDA activated current was not affected.

Although the NMDA receptor-ion channel is known to be permeable to Ca^{2+}, and therefore capable of activating CaM-kinase via autophosphorylation, there are very few studies describing the functional regulation of NMDA receptors by this kinase. In rat spinal dorsal horn neurons, the phosphatase inhibitor microcystin-LR enhanced responses to NMDA, presumably by increasing the basal activity of CaM-kinase. Furthermore, internal dialysis with the α subunit of CaM-kinase also enhanced the responses to NMDA, whereas heat-inactivated CaM-kinase was inactive.

In rat dorsal horn neurons the tyrosine kinase inhibitors, including genistein or lavendustin A, reduced aspartate-or NMDA-activated responses, including the NMDA receptor-mediated increase in intracellular Ca^{2+}. The current-voltage relationship for NMDA currents indicated that genistein acted in a voltage insensitive manner. The inactive inhibitors daidzein and lavendustin B did not affect NMDA receptor-mediated responses. Internal perfusion of the tyrosine kinase c-src enhanced the NMDA- activated responses, an effect that was prevented by prior immunoprecipitation or boiling of c-src.

NMDA receptor activation can also lead to tyrosine kinase activation. Stimulation of hippocampal neurons with glutamate resulted in Ca^{2+} influx and the transient tyrosine phosphorylation of a 39-kDa protein that is either identical to or closely related to microtubule-associated protein 2 kinase (MAP-2). This indicates that NMDA receptor activation may result in transduction of diverse signals after the initial well-characterized membrane conductance change.

The implications for synaptic plasticity and the role of NMDA receptors are still to be resolved. Despite the studies of protein kinases modulating NMDA receptor function, it was not until recently that serine/threonine or tyrosine protein phosphatases (PP) were considered to be potential regulatory agents. In hippocampal neurons, calyculin A, an inhibitor of PP1 and PP2A, or okadaic acid enhanced the whole-cell response to NMDA. Dose-response curve analyses revealed that only responses to higher concentrations of NMDA (approximately 20-30 μM) were enhanced by calyculin A, underlining the importance of examining the role of kinases and phosphatases by using more than one concentration of ligand. In cell-attached patches, calyculin A increased the probability of NMDA channel opening, whereas in inside-out patches, application of PP1 or PP2A reduced channel activity. The effect of PP2A, but not of PP1, was inhibited by concurrent application of 10 nM okadaic acid. Higher concentrations of okadaic acid inhibited both phosphatases. Support for the involvement of calcineurin in dephosphorylation and a potential physiological role was emphasized by studying the synaptic desensitization of NMDA receptors.

Repetitive activation of EPSCs in hippocampal neurons resulted in desensitization of NMDA receptors, an effect prevented by internally dialysing either with high concentrations of BAPTA (20 mM) to control intracellular Ca^{2+} levels or with ATP-γ-S, suggesting a role for phosphorylation. Moreover, application of specific inhibitors of calcineurin, such as cyclosporin A, FK506, or calcineurin inhibitory peptide, also prevented desensitization. Interestingly, desensitization was not prevented by the phosphatase 1 and phosphatase 2A inhibitor calyculin A or by the tyrosine phosphatase inhibitor sodium vanadate. Given the close localization of glutamate receptors with calcineurin and CaM-kinase, it would be of interest to investigate synaptic desensitization after intracellular dialysis with CaM-kinase.

Although vanadate appeared to have no effect on NMDA receptor synaptic desensitization, protein tyrosine phosphatases are able to modulate NMDA receptor function. Inclusion of sodium vanadate in the patch pipette resulted in enhanced NMDA-activated currents recorded from dorsal horn neurons, suggesting that receptor function is either directly or indirectly affected by tyrosine kinase-induced phosphorylation. Activation of protein kinase C by either phorbol esters or stimulation of metabotropic glutamate receptors in rat brain mRNA-injected oocytes caused an enhancement in NMDA-activated responses. This effect wits not reproduced by α-phorbols and received no contribution from the oocyte's endogenous calcium-activated chloride current. Prior, transient application of quisqualate or *trans*-1-aminocyclopentane-1,3-dicarboxylic acid (*trans*-ACPD) also enhanced the responses to NMDA. This effect lasted for approximately 10 min after removal of the metabotropic glutamate receptor agonists. The kinase inhibitor staurosporine inhibited the enhancement by quisqualate or phorbol ester treatment.

Recombinant NMDA receptors also contain consensus sequences for phosphorylation by protein kinases. Initially, putative consensus sequences involving CaM-kinase or PKC were identified in the NMDAR1 (ζ1) and NMDAR2A-NMDAR2C (ε1-ε3) subunits. These sequences were selected on the basis that this part of the mature protein was likely to be intracellular, following the prototypic four transmembrane domain model of a ligandgated ion channel.

However, this model was challenged after phosphopeptide mapping revealed multiple sites for phosphorylation by PKC in NMDAR1. Most of these sites were locaiized to a single alternatively spliced exon in the C-terminal domain, which previously was thought to have an extracellular location. Site-directed mutagenesis of NMDAR1 serines 889,890, 896, and 897 to alanines reduced, but did not abolish, the level of phosphorylation. The residual phosphorylation did not occur in the C-terminal and perhaps might involve the presumed loop between TM3 and TM4. Given the potential for phosphorylation by PKC, the C-terminal tail may have an intracellular location. Moreover, alternative splicing of this region could control the extent of phosphorylation of the NMDAR1 subunit.

Homomeric ζ1 (NMDAR1) receptors were modulated by activation of PKC using PMA and resulted in enhanced responses to NMDA. Responses recorded from the smaller NMDAR1 subunit, resulting from alternative splicing of the C-terminus and designated as ζ1-2, were similarly modulated by PMA in a staurosporine-sensitive manner. Heteromeric NMDA receptors revealed an interesting diversity with respect to modulation by PKC with a dependence on the ε subunit. Receptors composed of ζ1ε1 or ζ1ε2 were positively modulated by PMA, whereas recombinants composed of ζ1ε3 or ζ1ε4 were unaffected. The carboxy terminal region of the ε2 subunit was implicated in PKC-induced modulation following an analysis of chimeras formed from ε2 and ε3 subunits. Exchanging the C-terminus of ε2 for ε3 in the heteromeric complex ζ1ε resulted in a loss of modulation of PMA.

Interestingly, exchanging only the amino acids between TM3 and TM4 of ε2 for ε3 (thought previously to be involved with phosphorylation) did not affect the modulation by PMA. Although direct phosphorylation of the NMDAR1 subunit has been demonstrated, the role of the ζ1C-terminus in functional modulation by PKC is questionable. Removal of all five serine and threonine residues in the alternatively spliced C-terminus of the ζ1-2 subunit, did not significantly affect the level of enhancement induced by PMA when compared to the wild-type ζ1-2 subunit.

However, this study did not demonstrate biochemically whether the mutations removed all the PKC-induced phosphorylation of the ζ1-2 Subunit.

Whether PKC can phosphorylate the C-terminus of ε1 and ε3 subunits has not yet been demonstrated. Perhaps exchanging ε2 for ε3 in heteromeric subunit combinations exerts a remote conformational effect on NMDA receptor phosphorylation sites, rather than simply deleting these sites completely.

Clearly, the sites of phosphorylation in ζ1 subunits will still be present after exchange of ε subunits. The functional effects of phosphorylation can be influenced by alternative splicing of the NMDAR1 gene that results in 8 different receptor isoforms. Diversity is created by splicing in and out of three different exons. The amino acids encoded by the exons are inserted either in the N-terminal or at two sites in the C-terminal domains of the receptor. Interestingly, either the *presence* of the N-terminal insert or the *absence* of both C-terminal inserts increased the potentiation of NMDA-activated currents by PMA, indicating that different NMDAR1 splice variants are differentially modulated by PKC.

Curiously, the splice variant with the shortest C-terminal domain exhibited the largest enhancement following exposure to PMA. In contrast, one splice variant with the longest C-terminal tail exhibited the smallest enhancement, once more calling into question the relevance of phosphorylation sites on the C-terminal domain. The authors concluded that the inclusion or deletion of N- or C-terminal inserts may have allosteric effects on the phosphorylation sites for PKC. The identification of the sites of PKC-induced phosphorylation have become even more obscure following the study by Sigel and colleagues. Two heteromeric NMDA receptor subunit combinations were used, featuring NMDAR2A and the NMDAR1 splice variants, NMDAR1A or NMDAR1C.

The NMDAR1A splice variant contained both the C-terminal inserts, whereas the 1C version possessed only a single insert, nearest the C-terminal. Neither variant included the N-terminal insert. Application of PMA to oocytes expressing the recombinant NMDA receptors enhanced the responses to NMDA, which reached a peak after 10-15 min and eventually declined to less enhanced amplitudes. Staurosporine inhibited the enhancement and this effect was not reproduced by the inactive α-PMA. Larger enhancements were observed with NMDAR1C/NMDAR2A compared to NMDAR1A/NMDAR2A combinations. This is in accordance with the findings of Durand and colleagues, since those heteromeric NMDA receptors incorporating the splice variant with the shorter C-terminal tail (NMDAR1C), exhibited the largest enhancement to PMA.

Exchange of NMDAR1A for NMDAR1C results in the loss of at least four potential sites for phosphorylation by PKC, yet this leads to greater enhancement, which brings into question the involvement of the C-terminus in this phenomenon. Additional potential sites for phosphorylation between TM3 and TM4 in NMDAR1C and NMDAR2A were mutated to alanines (S747A in NMDAR2A; S658A, S748A, and T773A in NMDAR1C). The expression of different combinations of wild-type or mutant NMDAR1C/NMDAR2A subunits, resulting in receptors with up to four of the phosphorylation sites now mutated, were without effect on the enhancement caused by PMA. The sites between TM3 and TM4 were not functionally linked to the NMDA channel. Thus, to date, it is unclear which sites are involved in the PKC-induced modulation of the NMDA receptors.

Effects of ethanol on ion channels

Ethanol is more hydrophobic than water and in some instances can replace water within molecular structures, but it does not have the hydrogen bonding capacity of water. Ethanol readily enters molecular sites within multimeric ion channels, modifying inter-molecular forces and bonds that are important for the open-close-inactivation kinetic properties of channels.

The diversity of channel composition stemming from the multimeric structure results in subtypes of channels that have a spectrum of sensitivity to ethanol; this translates to regional differences in ethanol sensitivity in the brain, arising in part from differences in ion channel subunit composition within brain areas. Thus, many of the actions of ethanol can likely be ascribed to effects directly on ion channels. Ethanol affects both receptor-activated ion channels and voltage-gated ion channels. This chapter will review the structure and properties of various ion channels as well as the effect of ethanol on the channels. The two channels studied in great detail are the glutamate-NMDA receptor and the $GABA_A$ receptor. The acute intoxicating and incoordinating effects of ethanol are likely related to inhibition of subtypes of NMDA glutamate receptors and potentiation of certain subtypes of $GABA_A$ receptor ion channels.

Because glutamate is the major excitatory transmitter in brain and GABA the major inhibitory transmitter, the respective ion channel receptors are widely distributed and are among the most abundant receptor-operated ion channels in the central nervous system. These receptor ion channel complexes are not homogeneous but are a class of isoreceptors formed from various combinations of a variety of subunits. As the dose and, therefore, brain concentration of ethanol increase, the spectrum of ion channels altered broadens, and the impact on those channels most sensitive to ethanol is enhanced. The effects of ethanol on glycinergic, nicotinic cholinergic, serotonergic, as well as on voltage- gated Ca^{2+} and K^+

channels are also reviewed. These and other ion channels likely contribute to many of the acute actions of ethanol. Many studies have found effects on ion channel properties at ethanol concentrations consistent with mediating the euphoric and intoxicating effects of ethanol.

Furthermore, potent drugs that act specifically on ion channels can reverse or enhance the actions of ethanol consistent with ethanol directly modifying ion channel function. The effects of chronic ethanol consumption are also summarized. Alterations in NMDA, $GABA_A$, and voltage-gated Ca^{2+} channels are suggested to contribute to the development of ethanol tolerance, dependence, and withdrawal. Many of the other channels described have not been studied in this context.

Actions of ethanol on NMDA receptor function

Ethanol antagonizes NMDA-induced responses came from electrophysiological studies as well as from *in vitro* biochemical investigations showing that ethanol inhibits Ca^{2+} flux and NMDA-activated cyclic GMP. Electrophysiological studies in vivo have provided additional evidence that low to moderate concentrations of ethanol antagonize responses to NMDA. Recently, rats trained to discriminate doses of ethanol (1.0, 1.5, and 2.0 g/kg) from water, were found to substitute the NMDA antagonists phencyclidine (PCP) and MK-801 (dizocipline) for higher doses of ethanol.

In contrast, the competitive NMDA receptor antagonist CPPene did not substitute for ethanol. It was concluded that antagonism of NMDA receptor function by a channel-blocking NMDA antagonist affects the animals similarly to alcohol. This latter finding indicates that the inhibition of NMDA receptors is an important aspect of the action of ethanol on brain function. Additionally, it has been reported that an NMDA antagonist will prevent alcohol preference. Thus, there is indisputable evidence that moderate concentrations of ethanol antagonize NMDA responses, with functional consequences.

After the seminal findings concerning ethanol antagonism of NMDA Simson *et al.* using *in vivo* recording techniques in anesthetized rats, demonstrated that systemically administered ethanol antagonized the effect of NMDA on some, but not all, cells in the medial septum. Local application of ethanol onto neurons also reduced NMDA responses, eliminating the possibility that systemic administration of ethanol was acting at distant brain sites to produce inhibition of NMDA at the recording site. Furthermore, it has been established that the differential action of ethanol on responses to NMDA observed in the medial septum in anesthetized rats is also observed in rats given no anesthetic, indicating that the urethane anesthesia used in the previous studies was not responsible for the selectivity of ethanol for NMDA responses.

Initial extracellular recordings found sensitive and insensitive NMDA responses in the medial septum and substantia nigra reticulata (SNR). As in the medial septum and SNR, ethanol also inhibited the NMDA-induced increases in firing rate for some, but not all, spontaneously active neurons in the ventral tegmental area, the deep mesencephalic nucleus, and the red nucleus. In the hippocampus, ethanol inhibited NMDA responses from all neurons. However, it was not found to be active against NMDA responses in the lateral septum, suggesting a degree of regional specificity for inhibition. By measuring Ca^{2+} flux, A Wilson et at also discovered regional differences in the brain for ethanol inhibition of NMDA-stimulated responses, data consistent with earlier findings. Also, at 25°C ethanol antagonized responses to NMDA in slices from the hippocampus, but not in the caudate.

These data suggest that regional differences in the brain in NMDA receptor subtypes reflect differences in NMDA subtype sensitivity to inhibition by ethanol. Lovinger reported a relationship between the abilities of ifenprodil and ethanol to block NMDA responses. Over a 4-week period, the sensitivity of neocortical neurons in culture to both ethanol- and ifenprodil-induced inhibition of NMDA responses decreased, and both drugs were found to inhibit recombinant NMDA receptors containing the NMDAR2B subunit.

More recently, in vivo studies compared inhibition of NMDA responses by ethanol or ifenprodil in the same neurons. In the several brain regions investigated, the actions of ethanol and ifenprodil in antagonizing the neuronal NMDA response were similar. However, in some neurons ifenprodil antagonized NMDA but ethanol did not.

Mechanism of the differential inhibition of ethanol in some brain regions but not others appears to be related to NMDA subtypes, that is, NMDA receptors with different subunit composition. Alternatively, other synaptic inputs that alter NMDA responses, such as potentiation of GABA inhibition, could influence the NMDA response. Differential NMDA subunit composition in brain appears to be the most likely explanation.

Ethanol antagonizes NMDA in the CA-1 region of the hippocampus, a brain site where ethanol has no effect on GABA responses. Furthermore, bicuculline antagonizes responses to GABA without having an effect on NMDA excitation or on ethanol inhibition of NMDA (unpublished data). In addition, local application of ethanol produces the same effect as systemically administered ethanol, eliminating influences from distant brain sites as a cause of the differential action of ethanol on NMDA responses in the various brain regions thus far investigated.

Ethanol inhibits NMDA responses from recombinant NMDA receptor subunits in which $GABA_A$ receptors would be absent. These data suggest that subtypes of NMDA receptors, with different subunit compositions, in different parts of the brain determine the effects of various doses of ethanol on NMDA excitatory responses, with some subtypes being sensitive and others insensitive or less sensitive.Several studies have suggested that ethanol does have a differential action on subtypes of NMDA receptors.

Ifenprodil most potently inhibits a subset of NMDA receptors with the NMDAR2B subunit and also strongly inhibits those NMDA receptors that are sensitive to ethanol inhibition either in cultured neurons or *in vivo*. That ethanol has selective actions on particular subtypes of NMDA receptors is also suggested by reconstitution studies. Several investigators have combined the NMDAR1A variant with the four NMDAR-2 subunits in *Xenopus* oocytes or HEK-293 cells to examine the action of ethanol on responses to NMDA.

Consistent with the neuronal studies suggesting that inhibition of NMDAR2B by ifenprodil predicts inhibition by ethanol, expression of the ε2 subunit (mouse NMDAR2B subunit) combined with the subunit (mouse NMDAR1A variant) in Xenopus oocytes is significantly inhibited by concentrations of ethanol less than 25 mM. Higher concentrations of ethanol also inhibited εl (mouse NMDAR2A subunit) combined with ζ1, but concentrations as high as 100 mM ethanol had little or no effect on the ε3 (mouse NMDAR2C subunit) or ζ1 (mouse NMDAR2D subunit) subunits combined with 0 (mouse NMDAR1A variant).

Similar studies by Buller *et al.* found both NMDAR2A and NMDAR2B in combination with NMDAR1A to be the NMDA receptor subtypes most sensitive to inhibition by ethanol. There is consensus that the recombinant NMDA receptors consisting of either the NMDAR2C or the NMDAR2D subunit in combination with the NMDAR1A subunit are not sensitive to ethanol inhibition. Lovinger found that NMDAR1A/NMDAR-2A receptor was inhibited by ethanol; however, when NMDAR-2B was added, the recombinant NMDA receptor was considerably more sensitive to ethanol inhibition than was the recombinant receptor containing only the NMDAR2A subunit.

In vivo, NMDA receptors in the cerebral cortex likely contain combinations of NMDARI, NMDAR2A, and NMDAR2B subunits. The data, in entirety, suggest that NMDA receptors containing NMDAR2B subunits alone or in combination with NMDAR2A subunits are likely to be the subtypes of NMDA receptors that are most sensitive to ethanol inhibition, whereas those receptors containing NMDAR2C or NMDAR2D sub units are likely to be relatively insensitive to inhibition by ethanol. The presence of specific NMDAR1 receptor subunit variants may also contribute to differential sensitivity of NMDA subtypes to inhibition by ethanol. Koltchine *et al.* demonstrated that four of the NMDARI variants alone had differential sensitivity to ethanol, with the variant containing both the N-terminal and another insert being the most sensitive to inhibition when expressed in oocytes as homomeric assemblies.

Since NMDAR1A variants that lack an *N*-terminal insert are sensitive to ethanol when combined with the NMDAR-2 subunit, it is possible that NMDARI variants combined with various NMDAR2 variants constitute a spectrum of NMDA receptor subtypes with different sensitivities to inhibition by ethanol. When the NMDARI

B subunit that contains the *N*-terminal cassette was combined with the NMDAR2B subunit, the resultant receptor, as well as another receptor formed with NMDAR1B and NMDAR2A subunits were found to be more sensitive to ethanol inhibition of NMDA than were those receptors containing NMDAR2C or NMDAR2D subunits.

Gallagher *et al.*, combining the NMDAR1B variant with the NMDAR2B subunit, found that the recombinant receptor would bind ifenprodil. Thus, for the NMDARI variants investigated, an interaction of both the NMDARI splice variant with the various NMDAR2 subunits appears to modulate the sensitivity to ethanol; the NMDAR1 variants contain the inserts and the NMDAR2B subunits apparently are the most sensitive to inhibition by ethanol. Another aspect regarding differential inhibition of NMDA receptors by ethanol could involve an interaction with specific modulatory sites on the NMDA receptor.Glycine reportedly reverses ethanol inhibition of NMDA-induced Ca^{2+} flux and cGMP in cerebellar granule cells.

However, glycine did not reverse ethanol inhibition of NMDA-stimulated excitotoxicity in cerebral cortical neurons or patchclamp responses in cultured hippocampal neurons. Recent work by Buller *et al.* may explain these contradictory results. These workers demonstrated that responses to NMDA in Xenopus oocytes containing the recombinant NMDAR1A/NMDAR2B receptor were inhibited by the addition of ethanol at subsaturating concentrations of glycine. Inhibition was reversed by increasing the concentration of glycine, and at saturating concentrations of glycine, ethanol was less potent at inhibiting NMDA. However, other aspects of this study clearly demonstrated glycine-independent components. Thus, this work suggests that ethanol can have two distinct effects on responses to NMDA depending on the glycine concentration.

Furthermore ethanol does not appear to influence the polyamine site on NMDA receptors because polyamines do not compete for the ethanol inhibition of NMDA responses. Similarly, manipulations of various other modulatory sites (for example, Mg^{2+}, Zn^{2+}) or administration of the drug dithiothreitol (influencing oxidation and reduction) did not affect ethanol inhibition of NMDA in recombinant systems. Several studies have demonstrated that phosphorylation of the ionotropic glutamate receptors is another factor that might contribute to ethanol sensitivity.

The NMDA receptor subunits contain consensus sequences that allow for phosphorylation by protein kinase C (PKC) and perhaps by other kinases, as well as dephosphorylation by phosphatases. Snell *et al.* found that staurosporine and calphostin reversed the inhibitory of effect of ethanol on NMDA responses, suggesting that ethanol may facilitate phosphorylation of the NMDA receptor to inhibit responses to NMDA. Consequently, phosphorylation of the NMDA receptor may be required for ethanol and ifenprodil to inhibit NMDA. However, Tabakoff noted that PKC does not appear to have the same role in all cell populations in determining the effect of ethanol on NMDA responses.

Irrespective of a role for phosphorylation of receptors in the action of ethanol, this latter observation suggests that ethanol can act on more than one NMDA receptor subtype. Although ethanol clearly can act directly to inhibit NMDA receptor subtypes, the role of a secondary action via kinases is not well understood and adds an additional level of complexity to the brain regional differences in NMDA receptor sensitivity to inhibition by ethanol.

Most studies exploring the molecular mechanisms of ethanol-induced inhibition of NMDA receptors find ethanol to be a noncompetitive inhibitor of NMDA responses with a differential degree of efficacy. Rarely is any concentration of ethanol 100% effective at inhibiting maximal NMDA responses in reconstitution studies. Experiments corelating the hydrophobic aspects of a series of n-alcohols demonstrate that the potency of inhibition in cultured hippocampal neurons increases as the number of carbon atoms increases from one to five, reaches a maximum with six to eight carbon atoms, and then is abruptly lost, suggesting that a hydrophic pocket on the receptor protein may play a key role in the action of ethanol. In single-channel experiments using outside-out tear off patches, ethanol does not affect the conductance of the channel but produces a concentration-dependent reduction in the probability of the channel opening.

Actions of ethanol on kainate and AMPA receptors

Lovinger *et al.* reported that only high concentrations of ethanol (-100 mM) would antagonize responses to kainate from cultured neruons from fetal rodent brain. Similarly, ethanol inhibited responses from recombinant AMPA-type glutamate receptor subunits expressed in kidney 243 cells at concentrations greater than 100 mM. Weight *et al.* reported that in cultured neurons only ethanol concentrations greater than 50 nM affected the non-NMDA receptors, suggesting that ethanol's ability to produce general anesthesia could be attributed to inhibition of kainate and AMPA receptor function. High concentrations of ethanol (100 mM) inhibit kainate and AMPA responses in vivo in the rat locus coeruleus.

Nie *et al.* reported that ethanol reduced the amplitude of glutamate-induced excitatory postsynaptic potential (EPSP) in nucleus accumbens slices and attributed the change to an action oil non-NMDA receptors. Nonetheless, the glutamate receptor subtype would be in question for this latter study. Engberg and Hajos did not find an influence of ethanol on responses to non-NMDA receptor agonists. Ethanol inhibited the Ca^{2+} signal induced by quisqualate in mature cultured Purkinje neurons and, in oocytes expressing recombinant homomeric AMPA and kainate receptor subunits, the GLUR-3 kainate channel was inhibited to a greater degree than was the GLUR-6 receptor channel.

The inhibitory action of ethanol on GLUR-3 channel currents was enhanced in the presence of Ca^{2+}. A particularly important observation made by Dildy-Mayfield and Harris was that ethanol inhibition was enhanced at lower, compared to higher, concentrations of kainate. In fact, the inhibition of a low concentration of AMPA by ethanol wits like that seen for NMDA. The brain region may also determine whether ethanol affects non-NMDA receptors.

Concurring with this study, Snell *et al.* reported that the ability of ethanol to inhibit responses to glutamate in cultured cerebellar neurons depended oil the glutamate concentration, but the glutamate receptor subtype involved is not known. Martin *et al.* found that ethanol inhibited agonist responses to the non-NMDA receptors in hippocampal slices, but as noted in the earlier studies, this action depended greatly on the concentration of the agonist used.

Chandler *et al.* found-that AMPA-, but not kainate-, stimulated nitrous oxide formation in cerebral cortical cultures was inhibited by ethanol. Although there is no doubt that ethanol can inhibit responses of the non-NMDA receptors, the differences observed between the various studies regarding the concentration of ethanol required to affect responses to kainate or AMPA have to be explained. It is possible that the variable effects of ethanol relate to the differing subunit compositions of various non-NMDA receptor subtypes. Additionally, consensus sites are found on some subunits for protein kinase II and PKC, suggesting that phosphorylation could account for some of the differences observed.

Effects of ethanol on glutamate receptor subtypes

NMDA receptors may be involved in tolerance to ethanol. For example, when MK-801 (dizocilpine) was coadministered with ethanol, tolerance to ethanol was eliminated.

More recently, Khanna *et al.* found that inhibition of nitric oxide also suppressed the development of tolerance to ethanol; NMDA receptors are closely coupled to nitric oxide formation. The same group reported that rats given an NMDA antagonist before testing exhibited tolerance, whereas tolerance was not observed when the antagonist was given after behavioural testing.

Consequently, it was concluded that the alterations in NMDA function related to tolerance to ethanol may be dependent upon a learning component based upon cues by which tolerance is tested. Hyperexcitability of the central nervous system (CNS) is a key component of ethanol withdrawal. Both a reduction in GABA-mediated inhibition and a supersensitive response to glutamate appear to be involved. One of the earliest findings suggesting glutamate involvement was the finding that [^{3}H]glutamate binding is increased in the hippocampus of alcoholics.

Although the subtype of glutamate receptor is not clear this observation is consistent with increased glutamate receptor density. Several studies in neuronal cells in culture have indicated that a few days of chronic ethanol treatment leads to supersensitive NMDA-stimulated Ca^{2+} flux, as well as NMDA-stimulated excitotoxicity and NMDA-stimulated nitric oxide formation. Nitric oxide has been implicated in toxicity because of the formation of highly oxidative metabolites. Although both nitric oxide and excitotoxicity show supersensitivity because of shifts to the left in the dose-response curves, the finding that nitric oxide formation is maximal at the threshold concentration for NMDA-stimulated excitotoxicity suggests that nitric oxide may contribute to the excitotoxicity but is not the only factor involved.

Although supersensitivity was found in both of these responses, changes in the amount of NMDAR1 immunoreactivity were not found, suggesting the changes in the NMDA receptor structure may be responsible for the supersensitivity. Although the mechanisms are not totally resolved, chronic ethanol treatment can clearly induce NMDA supersensitivity.

Supersensitive NMDA responses could probably contribute to the hyperexcitability and seizures associated with ethanol withdrawal, and also cause neurotoxicity. Supersensitivity could occur through a number of mechanisms including changes in subunit composition or increased density of the NMDA receptor ion channel complexes. Increases in the density of NMDA receptors in C57BL mice treated chronically with a 7% ethanol liquid diet have been reported. After 7 days of chronic ethanol treatment [^{3}H]MK-801 binding in hippocampal membranes increased by approximately 16%. These animals were dependent upon ethanol as indicated by withdrawal seizures. Ail autoradiographic study also reported increased [^{3}H]MK-801 binding in cortex, hippocampus, and striatum that returned to control values in approximately 24 h, a time course similar to that for the return of seizure scores to control levels. An extension of these experiments with membrane binding studies found changes only in hippocampus and not cerebral cortex. [^{3}H]MK-801 and NMDA-specific [^{3}H]glutamate binding slightly increased in hippocampus during chronic ethanol treatment, but there were no changes in binding of [^{3}H]glycine or [^{3}H]CGS19755, a competitive NMDA antagonist, in hippocampus. Nor were any changes in any ligand binding noted in cerebral cortex.

Reconstitution studies demonstrated that NMDARlA subunits (the splice variant lacking the N-terminal insert but containing the C-terminal inserts) with NMDAR2A subunits could form receptors containing both glutamate antagonist and channel- blocking antagonist binding sites. Further studies combining NDAR1A subunits with NMDAR2B subunits resulted in different properties, particularly modulation of the channel by polyamines and high-affinity ifenprodil binding.

Thus, changes in some NMDA ligand-binding sites may be involved in subunit composition changes rather than an increased density of channels. Trevisan *et al.* found that a 12-week ethanol liquid diet in rats increased NMDARI immunoreactivity in the hippocampus, but not in cortex, striatum, or nucleus accumbens. Interestingly, a study of mRNA levels has indicated that chronic ethanol does not alter NMDARI mRNA level but increases NMDAR2A and NMDAR2B mRNA levels in hippocampus and cortex. Binding changes were not reported in this study. Since MK-801 apparently requires both NMDARl and NMDAR2 subunits for binding, an increase in binding could stem from changes in channel subunits without an increase in the density of channels.

Other studies have not found increases in MK-801 binding following chronic ethanol treatment of mice or rats. These differences could be due to different ethanol treatment protocols or the responses of different strains;

of animals. Long-term treatment of rats with ethanol was required to increase NMDAR1 immunoreactivity in the ventral tegmental area, whereas a 1- or 6-week treatment with a chronic 5% ethanol liquid diet was not sufficient.

Another factor that may relate to the differing results of ethanol treatment is that stress and treatment with glucocorticoids; increase NMDA receptor binding in a manner similar to treatment with ethanol. Because ethanol is a stressor and also increases glucocorticoids, it is possible that stress-induced glucocorticoids play a role in ethanol-induced increases in NMDA receptor binding in brain. However, the supersensitivity observer *in vitro* by several groups cannot be related to stress-induced changes in NMDA receptor binding. In vitro studies by Ticku's group have found that increased NMDA-mediated calcium flux, [^{3}H]MK801 binding and expression of NMDAR2B subunits during chronic ethanol treatment is modulated by the levels of glutamine in the media.

Although the exact molecular processes require additional experimentation, a number of studies support the hypothesis that chronic ethanol treatment results in supersensitive NMDA receptors. Another measure that supports NMDA hyperexcitability during withdrawal is Fos-like immunoreactivity (Fos-LI), which provides insight into the specific regions of brain activated by withdrawal. During ethanol withdrawal Fos-LI was observed in specific cortical regions, lateral septum, amygada (central nucleus), locus coeruleus, paraventricular thalamus, and dorsal central gray, suggesting that these brain regions, more than others, are involved in the symptoms of withdrawal in the absence of seizures. When rats were challenged with air puff, the effect in most brain, regions was intensified, with Fos-LI also being noted in the paraventricular, nucleus of the hypothalamus, the site of corticotrophin releasing factor (CRF)-containing neurons.

These findings provide a basis for associating specific regions of brain to symptoms and allow testing of the degree to which NMDA and non-NMDA receptors contribute to these changes during withdrawal. In this regard, after systemic administration of NMDA certain patterns of expression of Fos-LI in brain are similar to those observed during withdrawal from chronic ethanol exposure, supporting a role of NMDA supersensitivity in ethanol withdrawal. Further evidence for a direct involvement of NMDA receptors in ethanol dependence is obtained from studies showing that the NMDA antagonists MK-801 and CGP-39551 reduced the symptoms and prevented seizures during withdrawal from chronic ethanol treatment.

These studies suggest that NMDA receptor supersensitivity contributes to ethanol dependence and withdrawal. Several studies have suggested that chronic ethanol may also modify AMPA receptor properties. Although a 12-week ethanol liquid diet did not increase GluRI or GluR2 levels in the hippocampus of rats, it did increase GluRI immunoreactivity in the ventral tegmental area and substantia nigra. Furthermore, Breese *et al.* noted significant increases in GluR2 and GluR3 subunit immunoreactivity in the hippocampus of postmortem tissue from patients with a history of alcohol abuse. There is thus accumulating evidence that AMPA and kainate receptors may also be involved in the development of tolerance and dependence to ethanol.

Alcohol withdrawal symptoms increased over the prolonged duration of alcohol abuse ultimately resulting in seizures in alcoholic patients. To explain this finding, a "kindling" process, similar to that described by Goddard *et al.*, was hypothesized to account for this progressive intensification of withdrawal symptoms in alcoholics. Subsequently, in further support of this kindling hypothesis, the actual number of detoxifications for alcohol abuse was found to be more predictive of withdrawal seizures than was ethanol intake. From a series of investigations to define the basis of withdrawal audiogenic seizures after chronic ethanol treatment, the inferior colliculus was found to play a central role in the increased susceptibility to seizure activity during withdrawal from chronic ethanol exposure.

Subsequently, McCown *et al.* identified a specific site in the inferior collicular cortex that coincided with poststimulus wild running that accompanies audiogenic seizures during ethanol withdrawal. Upon repeated stimulation of the inferior collicular cortex, the running seizure gradually increased until seizure activity spread into the forebrain. Because the seizure complex observed after repeated electrical stimulation resembled that seen with audiogenic seizure activity after withdrawal from chronic ethanol exposure, the view that withdrawal seizures were mediated through the inferior colliculus was reinforced. Early studies that have examined the effect of prolonged chronic ethanol treatment or repeated withdrawals from chronic ethanol on symptoms and seizures did not provide an explanation for the worsening symptoms and seizures in alcoholics. Subsequently, taking advantage of the pivotal role noted for the inferior colliculus in ethanol withdrawal symptoms and the fact that a kindling phenomenon was associated with stimulation of this brain site, McCown and Breese re-evaluated the kindling hypothesis of Ballenger and Post.

Multiple withdrawals from chronic ethanol treatment were shown to reduce the number of stimulations required to attain inferior collicular kindling. In contrast, the number of stimulations required to reach kindling in amygdala was increased in rats on an ethanol diet over that for control diet rats. Although NMDA receptors are involved in withdrawal symptoms, it is not known whether NMDA receptor activation contributes to the kindling of the inferior colliculus caused by repeated withdrawals from chronic ethanol exposure. In addition to seizure activity, Daryanani *et al.* have evidence that the number of withdrawal episodes for an alcoholic patient is predictive of cerebral atrophy. The next section discusses the neurodegenerative disorders in alcoholics and the potential role of glutamate receptor activation in the pathological changes that correlate with withdrawal. Although withdrawal from chronic alcohol can result in acute CNS hyperactivity, the persistence of alcohol intake with inadequate nutrition and multiple episodes of withdrawal call result in severe persistent changes in brain function, including neurodegeneration.

Computer tomography (CT) and magnetic resonance imaging (MRI) studies in chronic alcoholics have demonstrated reduced cerebral mass, enlarged ventricles, and cerebellar atrophy. Microscopic examination of tissue at autopsy has revealed a diffuse and patchy loss of neurons and degeneration of remaining neurons in the brains of alcoholics. Cerebellar vermal and cerebral cortical atrophy has been noted. Nonetheless, it has been difficult to separate some of these changes from age-related changes. Several specific neurological disorders are associated with chronic alcohol abuse.

Wernicke's encephalopathy is a common syndrome associated with alcoholism and is caused by a thiamine deficiency resulting most likely from improper nutrition or absorption of thiamine in alcoholics. In particular, those alcoholics that do develop Wernicke's encephalopathy appear to have a decrease in transketolase activity, mostly due to a genetic abnormality in the enzyme. The triad of clinical symptoms includes encephalopathy, opthalnoplegia, and ataxia. Atrophy of the mammilary bodies is highly specific for this syndrome, but a large number of brain regions can be affected.

The mechanism of the neurodegeneration in Wernicke's syndrome is unknown, but excitotoxicity from glutamate has been linked to the neural destructive process. In animal studies, extracellular concentrations of glutamate in brain increase several fold during seizures in thiamine-deficient animals. Furthermore, MK-801, an NMDA antagonist, reduces experimental neurobiological symptoms and the severity of neural lesions in thiamine-deficient rats. It is not known whether coadministration of ethanol would enhance the degree of neurodegeneration seen in experimental Wernicke's encephalopathy (i.e., in thiamine-deficient animals).

A complicating factor of Wernicke's encephalopathy is Korsakoff's syndrome, in which there is a major loss of memory. The memory dysfunction correlates best with the presence of lesions in the thalamus. It must be recognized that Korsakoffs syndrome can occur in the absence of ethanol; however, in susceptible individuals, chronic ethanol may facilitate the course of the disease. This possibility has yet to be evaluated fully.

Other degenerative disorders associated with alcoholism include cerebellar degeneration, heptocerebral degeneration (associated with liver dysfunction), Marchiafava-Begnami syndrome, and central pontine myelinosis (also associated with liver malfunction). Glutamate receptor supersensitivity has been proposed to contribute to the degenerative processes associated with prolonged persistent exposure to ethanol and the hyperexcitable withdrawal syndrome in chronic alcoholism.

Most of the attention has been directed to the neurotoxic effect of the elevated intracellular Ca^{2+} presumably increased by NMDA. However, the recent recognition that selected non-NMDA receptors are also Ca^{2+} permeable and that extracellular Ca^{2+} is required for neurotoxicity of the non-NMDA agonists imply that glutamate receptor subtypes are also likely to contribute to neuronal toxicity associated with chronic alcohol exposure by a mechanism similar to that of NMDA. Furthermore, local administration of kainate can destroy neurons, and kainate has been widely used for this purpose.

Nicotinic cholinergic receptors allow movement of Ca^{2+} through the ion channel, and thus, this channel may also relate to neurotoxicity. In any case there is strong evidence that ethanol withdrawal hyperexcitability is related at least in part to NMDA supersensitivity and that this supersensitivity could underlie ethanol-induced brain damage. Ethanol inhibits glutamate NMDA receptor ion channels to a greater extent than it does the non-NMDA glutamate ion channel receptors. The heterogeneity of ion channel subunits and structures leads to differential sensitivity to ethanol. The NMDAR1/ NMDAR2B subunit containing NMDA receptors that bind ifenprodil display a high sensitivity to ethanol, whereas those containing NMDAR2C and NMDAR2D subunits appear to be less sensitive. The role of phosphorylation and multimers of other subunits requires further experimentation.

Ethanol interactions with $GABA_A$ receptors

γ-Aminobutyric acid (GABA) is the most ubiquitous inhibitory neurotransmitter in the brain that interacts with a family of receptors containing recognition sites for the anxiolytic and sedative benzodiazepines, barbiturates, and endogenous neurosteroids. These binding sites are linked allosterically to a GABA recognition site, and each site is involved directly or indirectly in the gating properties of integral Cl^- channels. GABA receptor mediated activation of Cl^- conductance results in membrane hyperpolarization and decreased neuronal excitability. Ethanol alters the gating properties of this receptor complex but has little or no binding affinity for the recognition sites for GABA, benzodiazepines, barbiturates, and cage convulsants on $GABA_A$ receptors.

GABA- mimetic drugs enhance and prolong the behavioural effect of ethanol, whereas antagonists shorten ethanol narcosis. Likewise, the benzodiazepines and GABA- mimetic drugs ameliorate the symptoms of ethanol withdrawal, whereas GABA antagonists potentiate these symptoms. Furthermore, benzodiazepine receptor inverse agonists such as RoJ5-4513 and FG-7142 antagonize many ethanol-induced behaviours in the rat, including intoxication. The behavioural effects of alcohols of different chain lengths have also been used to establish a link with alcohol interactions with GABA receptors. Recent studies have shown that whereas short-chain alcohols activate $GABA_A$ receptors, long-chain alcohols (more than 10 carbon atoms) no longer activate $GABA_A$ receptors. This loss of activity correlates with the effects of alcohols of different chain lengths on the loss of righting reflex in tadpoles, in which the distribution of alcohols in the brain has been measured.

Site-specific effects of ethanol on $GABA_A$ receptors that result in sedation have been demonstrated in the medial septum. Furthermore, seizures induced by ethanol dependence involve GABA receptors in the inferior colliculus, and chronic ethanol exposure alters several measures of GABA, receptor function that correlate with behavioural tolerance to ethanol. Direct evidence for interaction of ethanol with $GABA_A$ receptors at pharmacologically relevant concentrations has been obtained using subcellular brain preparations and cultured embryonic neurons in which ethanol and other short-chain alcohols have been shown to stimulate or potentiate $GABA_A$ receptor-mediated $^{36}Cl^-$ uptake.

Electrophysiological studies have confirmed that ethanol enhances $GABA_A$ receptor-mediated Cl- conductance, but only in specific brain regions or cell populations. The regional specificity of ethanol interactions with GABA was first observed by Givens and Breese who demonstrated that ethanol enhances GABA responses in the medial septal nucleus but not in the lateral septal nucleus. Subsequent studies have defined several brain regions in which electrophysiological responses to GABA are both sensitive and insensitive to ethanol.

Ethanol potentiation of $GABA_A$ receptors may be limited to very specific subtypes of $GABA_A$ receptors that have a unique regional distribution in brain. Criswell *et al.*, recently found that ethanol enhancement of' GABA responses in individual neurons can be predicted by the response of that neuron to zolpidem. Other $GABA_A$ isoreceptors may exhibit ethanol sensitivity under certain circumstances. Several investigators have demonstrated that ethanol modulates GABA responses in cerebellar Purkinje cells.

Some GABA responses are enhanced by ethanol in these cells if β-adrenergic receptors are coactivated; other GABA responses are antagonized in these cells. Cerebellar granule cells contain recognition sites for the benzodiazepine inverse agonist Ro15-4513 that antagonizes the effects of ethanol on righting reflexes as well as the effect of ethanol on GABA-mediated responses. Since this site is not sensitive to zolpidem, it is assumed that it represents a distinct $GABA_A$ is receptor that responds to ethanol. The subunit composition of this receptor is likely to include α_6 subunits because this subunit appears to contain the Rol 54513 recognition site. In cerebral cortex, Ro15-4513 labels a $GABA_A$ receptor consisting of $\alpha 4$ subunits.

Chronic ethanol administration increases, as well as binding to α_4 and α_6 subunits, respectively. Therefore, these $GABA_A$ receptor subtypes probably exhibit ethanol sensitivity in brain. The question of whether ethanol enhancement of GABA responses is dependent on receptor subunit composition has been directly addressed in various recombinant expression systems. In frog oocytes, the expression of the human γ_{2L} subunit has been reported to be required for ethanol potentiation of GABA responses, but other studies have not confirmed this effect. In mouse LTK cells, recombinant receptors containing either $\alpha_1\beta_1\gamma_{2L}$or $\alpha_1\beta_1\gamma_{2s}$ subunits are sensitive to ethanol, although greater sensitivity was observed when the γ_{2L} subunit was expressed. In the HEK293 expression system, the γ_2 Splice variant does not influence ethanol enhancement of GABA responses. Conversely, stably transfected PA3 cells are only sensitive to ethanol when the γ_{2L} variant is expressed with α_1 and β_1 subunits. The lack of consistent results in recombinant expression systems suggests that factors other than subunit composition can influence ethanol interactions with $GABA_A$ receptors. These factors could include differences in membrane properties, intracellular messengers, or receptor assembly that may exist among recombinant expression systems and mammalian brain. The extent to which these factors modulate the functional properties of GABAA receptors is of considerable importance in the, interpretation of structure-function studies in recombinant expression systems.

Ethanol shares several pharmacologic actions including anxiolytic and sedative activities with barbiturates and benzodiazepines, and under appropriate circumstances, cross-tolerance or cross-dependence may develop. The similarities of the actions of ethanol with those of benzodiazepines and barbiturates further suggest that all three drugs may share some mechanism(s) of action. Tolerance to the sedative and intoxicating effects of ethanol has been postulated to result from a compensatory decrease in GABA-mediated inhibition in the brain.

Alterations in the function of the $GABA_A$ receptor Cl^- channels contribute to some of the signs and symptoms of the ethanol withdrawal syndrome. Chronic exposure of rats to ethanol produces physical dependence and tolerance. These effects are associated with a decrease in the sensitivity of the $GABA_A$ receptor-coupled Cl^- channel in the cerebral cortex of dependent animals when blood ethanol levels are greater than 150 mg/dl. Muscimol-stimulated ^{36}Cl uptake was decreased by 26% and pen tobarbital-stimulated $^{36}Cl^-$ uptake was decreased by 25% after chronic ethanol inhalation.

The ability of ethanol (20 nM) to potentiate muscimol-stimulated $^{36}Cl^-$ uptake was completely lost in rat cerebral cortical synaptoneurosomes and in mouse cerebellar microsacs following chronic ethanol administration. Benzodiazepine enhancement of mucimol-stimulated Cl^- flux is reduced in the cerebral cortex of mouse microsacs, whereas the functional efficacy of inverse agonists; is enhanced. Behavioural responses to injections of muscimol into the substantia nigra and to subcutaneous THIP injections are also reduced after chronic ethanol administration.

Similarly, chronic ethanol administration reduces electrophysiological responses to GABA in the medial septal nucleus and GABA-induced Cl^- flux in spinal cord cultured neurons. Following the completion of ethanol withdrawal, the decrements in muscimcl-stimulated $^{36}Cl^-$

uptake in rat cerebral cortical synaptoneurosomes and flunitrazepam potentiation in mouse cortical microsacs are completely reversed, as would be predicted if these neurochemical changes are related to the behavioural state of withdrawal. Alterations in the density or affinity of brain $GABA_A$/benzodiazepine receptors following chronic ethanol administration have yielded conflicting results and cannot explain the development of tolerance or withdrawal.

For example, Ticku and Burch, as well as Unwin and Taberner have reported a decrease in the density of low-affinity [^{3}H]agonist binding sites in brain after chronic ethanol exposure to rats or mice. Devries *et al.* have reported a reduction in the ability of GABA to enhance [^{3}H]flunitrazepam binding in brain membranes prepared from ethanoltreated mice. However, others have failed to find alterations in the number or affinity of $GABA_A$ receptors, Chronic ethanol administration does not alter the density of [^{3}H]flunitrazepam or [^{35}S]TBPS binding sites in brain. Recently, Mhatre *et al.* reported an increase in the densityof specific binding sites for [^{3}H]Rol5-4513 in the ratcortex and cerebellum following chronic ethanol administration, as well as increased sensitivity to its behavioural effects. We have proposed that the alterations in $GABA_A$ receptor function observed following chronic ethanol administration are the result of a change in the synthesis and expression of subpopulation(s) of $GABA_A$ isoreceptors.

Chronic ethanol administration may have differential effects on the expression of individual $GABA_A$ receptor subunits, accounting for the diverse effects of ethanol on different radioligand binding sites. If ethanol alters the expression of specific populations of $GABA_A$ isoreceptors, radioligands that measure all GABA receptors may lack the selectivity to detect changes in one or more receptor Subunits or subpopulations of $GABA_A$ isoreceptors. Chronic ethanol administration differentially alters the expression of GABAA receptor subunit mRNAs in the cerebral cortex and cerebellum.

The levels of $GABA_A$ receptor α_1 subunit mRNAs are reduced whereas α_4 subunit mRNAs are increased by approximately equal amounts in cerebral cortex. In the cerebellum, likewise, decreases in $GABA_A$ receptor α_1 subunit mRNAs and increases in α_6 subunit mRNA levels are found. These changes in mRNA levels suggest alterations in the expression of' the corresponding proteins that could account for the alterations in receptor function and binding that have been observed. For example the increases in α_1 and α_6 subunit expression probably explain the increases in [^{3}H]Rol5-4513 binding and sensitivity following chronic ethanol administration. The increased expression of α_4 subunits may underlie the reduced sensitivity to GABA and benzodiazepine agonists because recombinant GABA receptors with $\alpha_4\beta_2\gamma_2$ subunits are less sensitive to GABA agonists; and benzodiazepines than $\alpha_1\beta_2\gamma_2$ receptors.

Ethanol-dependent and withdrawn rats are also sensitized to the effects of the neurosteroid 3α, 5α.-THP. This effect may be related to the increase in $\gamma_1$subunit mRNAs following chronic ethanol exposure because in recombinant expression studies γ_1 subunits convey greater neurosteroid sensitivity. The observation that ethanol tolerance is associated with changes in subunit composition also supports the hypothesis that subunit composition influences ethanol sensitivity.

Ethanol potentiation of $GABA_A$ receptor function is abolished at the cellular level in rats that are ethanol tolerant and dependent. The changes in subunit expression that result from chronic exposure to ethanol may contribute to this reduction in ethanol sensitivity.

However, alterations in the phosphorylation state of $GABA_A$ receptors may also be involved. Recent Studies in recombinant receptors, in which subunit expression is controlled by a dexamethazone-sensitive promoter, exhibit changes in receptor function that cannot be attributed to alterations in subunit expression. In addition, mouse mutants that lack a PKC isoform exhibit a loss of ethanol sensitivity both *in vivo* and at the cellular level.

Thus, ethanol tolerance and dependence may also involve second messenger effects on $GABA_A$ receptors. Dynamic changes in $GABA_A$ receptor subunit mRNAs are observed during ethanol withdrawal. $GABA_A$ receptor α_1, α_4, and γ_1 subunit mRNAs return to control levels, whereas β_2 and β_3 subunit mRNA levels increase compared to both controls and dependent rats. At this time point, corresponding changes in protein expression are unlikely because the receptor turnover rate ($t_{½}$) is estimated to be 1 to 2 days. Therefore, $GABA_A$ receptor polypeptide expression is probably similar to that found in ethanol- dependent rats.

Alterations in $GABA_A$ receptor subunit composition, in vivo, may result from adaptations to chronic drug exposure This could result in alterations in the functional properties of receptors with no change in the total number of receptors expressed. Hence, $GABA_A$ receptors may be regulated by alterations in subunit expression that result in the assembly of structurally and functionally distinct GABA isoreceptors. This mechanism appears to play a role following chronic exposure to ethanol and possibly other drugs that modulate $GABA_A$ receptors.

Thus, $GABA_A$ receptors are a major site of action of ethanol, likely contributing to many of the effects of ethanol. The heterogeneity of ion channel structure results in a differential sensitivity to ethanol with certain subunit combinations, such as the $\alpha_1\beta_x\gamma_2$ subunit being particularly sensitive to ethanol. Ethanol appears to have both direct effects on channel gating as well as effects secondary to changes in phosphorylation. The exact role of these various processes requires additional investigation.

$GABA_A$ receptor modification during chronic ethanol exposure also contributes to tolerance, dependence, and withdrawal from ethanol. A variety of studies have indicated that there is a change in receptor ion channel function during chroric ethanol exposure, and this appears to lead, to changes in subunit composition. Although a detailed and critical understanding of the role of $GABA_A$ receptors in acute and chronic responses to ethanol will require additional experimentation, it is clear that $GABA_A$ receptors are a major site of ethanol's actions.

Action of ethanol on glycine receptors

There are only a few reports showing that ethanol enhances the effect of glycine. Engblom and Akerman found that ethanol enhanced both GABA-and glycine-mediated Cl^- flux in synaptoneurosomes prepared from whole brain. Aguayo and Pancetti showed that a strychnine-sensitive glycine current from cultured hippocampal cells was enhanced by ethanol. In cultured spinal cord neurons, ethanol enhanced glycine-induced Cl^- currents to a much greater extent than it enhanced GABA-mediated Cl^- conductance. Whereas only 62% of spinal cord neurons exhibited ethanol potentiation of GABA, all of the spinal cord neurons exhibited ethanol potentiation of glycine-induced Cl^- conductance.

Because ethanol enhances glycine in the spinal cord and only the α_1 and β subunits have been found at, this site, the α_1,β isoform of the glycine receptor appears to be ethanol sensitive. Additionally, this isoform of the glycine receptor is the subunit combination that binds Strychnine with highest affinity. Therefore, it might be expected that all strychnine-sensitive glycine receptors will be sensitive to ethanol. However, other glycine isoreceptors bind strychnine, albeit with lower affinity. Additionally, the glycine receptor α_1 subunit exhibits alternative splicing at an exon encoding the cytoplasmic loop adjacent to transmembrane region M3. There is an eight-amino-acid insertion at this site that generates a novel Putative phosphorylation site on the α_1, subunit. Therefore, the expression of this α_1 subunit splice variant may also he important for ethanol sensitivity of glycine receptors.

In support of this concept, Criswell *et al.*, found that both ethanol sensitive and ethanol insensitive glycine receptors were blocked by strychnine. In rat brain, preliminary evidence suggests that ethanol augments respouses to glycine in some, but not all, cells of the medial septum, ventral pallidum and SNR. Additionally, Nestoros and Simson *et al.*, found that ethanol did not enhance the effect of glycine in the cortex or in the inferior collicular cortex, respectively. As noted above, ethanol enhanced responses to glycine in the medial septum, a site containing α_2 and β subunits, but with little or no strychnine binding. These data not only provide evidence for at least two types of glycine receptors-ethanol-sensitive and ethanol-insensitive subtypes-but also demonstrate that ethanol has a regionally specific action on responses to glycine in brain. Thus, the glycine receptor, like the other ligand-gated ion channels, shows heterogeneity in its response to ethanol.

The mechanism or mechanisms underlying this heterogeneity in the effect of ethanol on responses to glycine is unknown but may involve differential expression of glycine receptor subunits resulting in some cells containing an ethanol sensitive isoform of the glycine receptor, whereas others do not. Other likely sources of heterogeneity are differences in posttranslational processing or differential modulation by coagonists. For example, the degree to which agents such as Zn^{2+} modulate the glycine receptor has not been sufficiently studied. The importance of glycine receptor subtypes in brain to the behavioural action of ethanol will only become evident following further study of this neglected neurotransmitter system. The neuronal nicotinic cholinergic (nACh) receptor is another member of the superfamily of ligand-gated ion channels.

The easy accessibility to the muscle nACh receptor at the neuro-muscular junction and the ability to purify large amounts of the receptor protein from *Electraphorus* (eel) or *Torpedo* (ray) made this receptor- gated ion channel the model for biochemical and biophysical studies of all ion channels controlled by a neurotransmitter. However despite abundant data describing the neuromuscular nACh receptor, it has become clear that the neuromuscular nACh receptor is a relatively poor model for defining the specific functional aspects of neuronal nACh receptors.

The recent pharmacological definition of distinct nACh receptors in brain and the molecular biological studies that defined several neuronal nACh receptor subunits with distinct distributions in brain have provided an impetus for further studies, particularly because of the health risks from the use of tobacco and the interaction of ethanol and nicotine that perpetuates chronic abuse of these substances.

Interactions between ethanol and nicotine

Behavioural evidence for a relationship between ethanol and nicotine accrues from the finding that both drugs produce an anticonflict action in the elevated-plus maze and both are also active in the mirrored-chamber test. Similarly, ethanol preferring (P) rats are significantly more sensitive to the ethanol-like, discriminative properties of nicotine than are nonpreferring (NP) rats. Futhermore, chronic infusion of nicotine increases the preference for ethanol in rats. Ethanol increased locomotor activity, which was enhanced by low doses of nicotine and counteracted by mecylamine. Similarly, when ethanol alone did not stimulate locomotor activity in a rat strain, it enhanced the locomotor stimulation induced by nicotine.

Calabresi et al showed that nicotine excited neurons in the ventral tegmental area, which is the point of origin for dopaminergic neurons innervating the nucleus accumbens via the medial forebrain bundle (MFB). Microinjection of ethanol into the ventral tegmental area also acts as a reinforcer for operant response, presumably by a direct action on neurons in that site, providing a potential site of interaction between nicotine and ethanol. Previous efforts have implicated the ventral tegmental area and dopaminergic neurons in nucleus accumbens in the reinforcing properties of cocaine and amphetamine.

Ethanol also results in increased dopamine release in the nucleus accumbens, which can be blocked by administration of the nACh antagonist mecamylamine. Based upon the observation that ethanol and nicotine facilitated intracranial self-stimulation, an involvement of dopaminergic pathways was proposed to account for this behavioural change induced by ethanol and nicotine. Involvement of dopamine in the activation of self-stimulation by these drugs would seem a reasonable possibility, because dopamine release was shown nearly two decades ago to play a critical role in maintaining a self-stimulation response.

In addition to the action on the ventral tegmental area, nicotine increases the release of dopamine from striatal neurons. This action of nicotine is on presynaptic dopaminergic neurons because producing lesions of these neurons causes a dramatic reduction of ligand binding to nACh receptors in the striatum. However, the release of dopamine from striatal neurons by nicotine is antagonized by ethanol.

These studies provide evidence for a complex involvement of cholinergic-dopaminergic mechanisms in the actions of ethanol. It is important that because nicotine, the major constituent in tobacco, and ethanol both act on nACh receptors, the functional and behavioural interaction of these abused substances should continue to receive attention; the direct interactions of nicotine and ethanol oil neuronal nACh receptors could provide an explanation for why the majority of alcoholics smoke.

Mancillas *et al.* reported that ethanol enhanced the effect of acetylcholine in the hippocampus without affecting GABA-mediated inhibition. Beyond this observation on acetylcholine, electrophysiological studies of the action of ethanol on responses to specific nACh agonists on specific neuronal nACh receptors subtypes have been meager. Criswell *et al.* showed that ethanol influences responses to nicotine from some, but not all, neurons in the substantia nigra reticulata, proving that ethanol- sensitive and ethanol-insensitive nACh receptors existed.

Subsequently, ethanol we's found to block the inhibitory response to nicotine on some, but not all nACh receptors in the medial septum. Importantly, by evaluating the response to nicotine on medial septal neurons in the presence of Mg^{2+} nicotine was shown to act on presynaptic nACh receptors to facilitate the release of GABA, an interpretation congruous with other data indicating that nicotine can have a presynaptic action. It is known that nicotine- induced dopamine release from striatal synaptosomes is blocked by ethanol, whereas ethanol does not affect $^{86}Rb^+$ efflux induced by nicotine from thalamic synaptosomes. This latter result agrees with that of Yang *et al*. who found ethanol-sensitive and ethanol insensitive nACh receptor responses in brain.

These studies suggest differential responses of various nACh receptor subtypes as discussed for other channels. In addition to its action on presynaptic nACh receptors, ethanol has an action on postsynaptic responses to nicotine (the nicotine response not blocked by Mg^{2+}). In the cerebellum, ethanol blocked the nicotine-induced inhibitory response and enhance the excitatory response to nicotine.

Collectively, the results obtained in the medial septum and cerebellum, indicate that future studies within a given region of brain must evaluate whether responses to nicotine in vivo involve an action on pre-or postsymaptic nACh receptors. Brain regions in which ethanol has influenced nicotine responses include the cerebellum, SNR, ventral tegmental area, and hippocampus.

Since ethanol was administered locally in the majority of studies in which extracellular electrophysiological recording in vivo was used to assess responses to locally applied nicotine, the influence of ethanol on nACh function must certainly reflect a local action. The scarcity of investigations into the actions of ethanol on nACh receptors underscores the need for such studies to determine which effects of nicotine and ethanol related to presynaptic and postsynaptic nACh receptors, as well as which subsets of nACh receptors are sensitive to ethanol.

nACh receptors and cloning of several subunits for nACh receptors, there are clearly different nACh isoreceptors in brain. The view has been expressed that the heterogeneity of nACh receptor subtypes (derived from the molecular structure of the nACh isoreceptors) would be a necessary condition for the specific and variable effects of ethanol and nicotine in brain. However, the subunit components of the various presynaptic and postsynaptic endogenous neuronal nACh receptor subtypes present on neurons that are sensitive or insensitive to ethanol have not yet been identified.

Recent work from transient expression of the α_7 nACh receptor in oocytes indicates that ethanol will act on neurons containing the α_7 nACh receptor subunit. Thus, it is possible that α–BTX or MLA could be used to identify ethanol-sensitive receptors containing the α_7 subunit. In this regard, α-BTX will antagonize most of the inhibitory responses in the medial septum, a site at which ethanol will antagonize the response to nicotine. Therefore, it could be suggested that the nACh receptor in the medial septum that is sensitive to ethanol contains the α_7 subunit. It is of interest that the majority of the excitatory nicotinic responses from interneurons in the cerebellum are also antagonized by α-BTX.

However while this nACh receptor would be anticipated to contain the α_7 subunit, ethanol enhances this response to nicotine, in contrast to the inhibition of nicotine by ethanol observed for the α-BTX-sensitive nACh receptor in the medial septum. These latter findings suggest that more than one Type I α-BTX-sensitive nACh receptor subtype exists—one in which ethanol antagonizes the response to nicotine and another in which ethanol enhances the response to nicotine.

Ethanol did not affect all excitatory responses from interneurons in the cerebellum, suggesting that not all neurons containing nACh receptors with the α_7 subunit are sensitive to ethanol. It is proposed that the α_7-containing nACh receptors on interneurons that are insensitive to ethanol enhancement of nicotine will contain a differing combination of nACh receptor subunits than those neurons sensitive to this action of ethanol on nicotine. deFiebre and Meyer have reported that responses from the $\alpha_4\beta_2$ subunit combination expressed in *Xenopus* oocytes is not sensitive to ethanol. This observation would suggest the possibility that, if α-BTX or MLA does not block the response to nicotine, DHβE blockade would predict the in vivo presence of a nACh is receptor containing the $\alpha_4\beta_2$ subunits that is insensitive to ethanol.

In addition, Collins has speculated that the nACh receptor responsible for dopamine release from striatal synaptosomes contains the α_3 nACh subunit. This nACh receptor in the striatum is known to be sensitive to ethanol. Based upon these background data, it is anticipated that multiple nACh isoreceptors will be identified showing sensitivity to ethanol. Similarly, nACh receptors that do not show sensitivity to ethanol will be identified, one of which presumably will be $\alpha_4\beta_2$. As implied earlier, this molecular explanation for the action of ethanol on different nACh receptors is analogous to, the one that has been used to explain the variable effect of ethanol on the function of different $GABA_A$ and NMDA receptors.

Chronic nicotine administration upregulates nACh receptor binding, even though there is a tolerance to the actions of nicotine. Yang and Buccafusco demonstrated that chronic blockade of nACh receptor with DHβE caused an elevation of [^{3}H]cytisine binding but produced no tolerance to acute agonist, suggesting that increased levels of binding alone do not account for tolerance. It would thus be expected that the tolerance exhibited after chronic nicotine treatment would be related to a desensitized state by nACh receptors, but this has not been established. Sanderson found the upregulation, of nicotinic receptors after chronic nicotine exposure to be region specific, indicating that not all nACh receptors are affected equally by such chronic nicotine treatment.

There is cross-tolerance between certain physiological functions caused by chronic nicotine administration and those caused by chronic ethanol administration. Additionally, the degree of nicotine tolerance was found to differ between long- and short-sleep mice bred for differing responses to ethanol. Taken together, this series of investigations would point to a change in nACh receptor function after chronic ethanol treatment like that seen for nicotine; however, the generality of this conclusion concerning the chronic effect of ethanol on nACh receptors has not been evaluated electrophysiologically or with autoradiography of ligands that bind to the classified nACh receptor subtypes.

Because ethanol can alter desensitization of muscle nACh receptors, the chronic effect of ethanol on desensitization of neuronal nACh receptors should also be assessed. Regardless, because of the acute effects of ethanol on some, but not all, responses to nicotine, it can be anticipated that there will be a heterogeneity of effects on nACh receptors after chronic ethanol exposure, just as observed after acute ethanol administration and chronic nicotine exposure. Smoking results in the appearance of chronic low levels of nicotine in the plasma of smokers. Such low plasma levels of nicotine would presumably be sufficient to desensitize the majority of endogenous nACh receptors, although, as mentioned earlier, differences in the degree and rates of desensitization between differing isoforms of the nACh receptor apparently occur.

Desensitization of the nACh receptor by chronic nicotine administration has been hypothesized to account for the upregulation of [^{3}H]nicotine binding in both rats and human smokers. Chronic nicotine has been shown to upregulate both $\alpha_4\beta_2$ and α_7-containing isoreceptors. However, the $\alpha_4\beta_2$ nACh combination is not believed to be affected by acute ethanol. Nonetheless, upregulation of the α_7-containing isoform of the nACh receptor by chronic nicotin intake may be of special relevance to the interactions between smoking and drinking because this isoform is present on presynaptic receptor and ethanol has been shown to block the action of nicotine on α_7 subunit-containing nACh receptors.

Furthermore, the concentration of nicotine in the plasma of smokers acutely activates presynaptic nACh receptors that control release of glutamate, while having very little effect on postsyriaptic nACh receptors. Nonetheless, the degree to which chronic low levels of nicotine affect any given endogenous neuronal nACh receptor subtype, including presynaptic nACh receptors, is an unanswered question. In addition, it is not known how chronic low levels of nicotine influence the pharmacology of ethanol or how chronic repeated exposure to ethanol affects nACh receptors in the presence or absence of nicotine.

It is known whether cross-tolerance occurs behaviourally after chronic treatment of these addictive drugs. Consequently, it could be anticipated that ethanol would cause adaptive changes in nACh receptor function similar to those caused by nicotine, but this hypothesis has not been tested. However, an exception to this generalization may be the $\alpha_4\beta_2$ nACH receptor, as this subtype appears resistant to the action of ethanol. Such apparent discrepancies of logic emphasize the need to expand research on the acute and chronic actions of ethanol on nACh receptor function. Converging clinical data have shown that alcoholism and excessive drinking are several times more prevalent in smokers than in nonsmokers.

It is also known that the ability to treat alcoholism is enhanced if individuals stop smoking. Because of the interactive association of these abused substances in alcoholism, the challenge will be to resolve the basis of this strong interaction between the action of chronic ethanol and chronic nicotine exposure.Consequently, it is extremely important that we have a complete understanding of the direct interactions that these chronically abused substances have on various nACh isoreceptors in brain, both alone an in combination. The critical question to be answered is how changes in nACh receptor function, induced by the presence of chronic nicotine and ethanol, is contributing to and perpetuating alcoholism and the need to smoke.

In the section on glutamate receptors, the neural hyperexcitability associated with withdrawal from chronic ethanol exposure is believed to result in a loss of neurons caused by excitotoxicity. Because nACh receptors control Ca^{2+} flux, it is possible that these receptors contribute to neurotoxicity through a change in intracellular Ca^{2+} levels, the mechanism suggested for the neurotoxicity caused by glutamate release during withdrawal from chronic ethanol exposure. This possibility that alterations in nACh receptor sensitivity could contribute to neurotoxicity- appear to have been overlooked.

Choline acetyltransferase. (CAT) levels are reduced in brains of alcoholics, suggestive of a loss of cholinergic neurons. Additionally, cholinergic neurons in the nucleus basalis of the forebrain were found to be reduced in three patients with Korsakoffs syndrome. In well-nourished rats after long-term ethanol exposure, a loss of CAT and choline uptake, indicative of neuronal loss, was observed. Memory loss in these rats on long-term treatment with alcohol was corrected by transplantation of cholinergic neurons into the hippocampus and neocortex.

These data suggest that cholinergic projection neurons from the nucleus basalis are particularly vulnerable to the mechanisms that result in neurotoxicity after chronic ethanol exposure, but which neurotransmitters contribute to the neurotoxicity upon chronic ethanol exposure and withdrawal to these cholinergic neurons is not known at this time.

Effects of ethanol on voltage-gated Ca^{2+} channels

Ethanol inhibits depolarization-induced Ca^{2+} influx through voltage-gated Ca^{2+} channels in synaptosomes and presynaptic nerve terminals without altering basal uptake.

Ethanol also reduces voltage-dependent Ca^{2+} influx in cultured neuronal and PC12 cells. In these preparations, ethanol inhibits Ca^{2+} flux during the first few seconds of depolarization through Ca^{2+} channels and apparently does not involve other mechanisms such as Na^+/Ca^{2+} exchange that operates at later times and appears to be less sensitive to ethanol. Most studies have focused on the L-type voltage-gated Ca^{2+} channels. Further support for the involvement of L-type voltage-gated Ca^{2+} channels in ethanol's response comes from studies showing that dihydropyridines potentiate the acute pharmacological effects of ethanol, including ethanol-induced hypothermia, motor incoordination, and sedation; also the L-type channel agonist Bay K8644 decrease the anesthetic effects of ethanol.

Inhibition of L-type voltage-gated Ca^{2+} channels has also been related to ethanol-induced decreases in arginine vasopressin release, contributing to the diuretic effect of ethanol. Electrophysiological studies in PC12 cells have found that ethanol inhibited dihydropyridine-sensitive currents, namely L-type channels, in a dose-dependent manner, but did not inhibit ω-conotoxin-sensitive currents, namely N-type currents.

When undifferentiated PC12 cells were compared to NGF differentiated cells, the effect of ethanol was found to be significantly greater in the undifferentiated cells suggesting that subunit changes in L-type channels or other factors such as G proteins may regulate sensitivity to ethanol inhibition of L-type voltage-gated Ca^{2+} channels. Ethanol inhibition appears to involve a shift in steady state inactivation, decreasing the open time. These studies suggest that ethanol inhibits voltage-gated Ca^{2+} channels, with certain subtypes of the L-type channels being the most sensitive. Although many flux Studies in intact brain have used fairly high concentrations of, ethanol (100 mM) in order to obtain strong effects, electrophysiological studies in some systems have observed much more potent actions of ethanol (10 mM) on L-type voltage-gated Ca^{+} channels, suggesting that subtypes of these channels may be particularly sensitive to ethanol. Differences between brain regions, and responses in different preparations are likely the result of different subunit composition of voltage-gated Ca^{2+} channels.

Variety of studies suggest a role for voltage-sensitive Ca^{2+} channel the effects of regulation in ethanol dependence, particularly studies showing that dihydropyridines, which block the L-type Ca^{2+} ethanol withdrawal in mice and rats. Extracellular recordings made from mouse hippocampal slices prepared after chronic treatment with ethanol alone or ethanol and nitrendipine suggest that the electrophysiological changes that Occur during ethanol withdrawal are also prevented by L-type Ca^{2+} channel blockers.

Furthermore, Bay K 8644, and L-type Ca^{2+} channel activator, increases withdrawal-induced hyperexcitability in mouse hippocampal slices prepared after chronic ethanol treatment, whereas PN200-110, an antagonist, decreases withdrawal-induced hyperexcitability. Several studies have indicated that chronic ethanol treatment increases dihydrophyridine binding sites in membranes prepared from whole brain and cerebral cortex from mouse and rats.

Chronic ethanol administration also increases the functional effects of Bay K 8644—inositol lipid breakdown and neurotransmitter release. These effects on dihydropyridine binding and function are blocked by Ca^{2+} channel antagonists. Clearly, regulation of dihydropyridine binding sites are important in the development of alcohol dependence and the manifestations of alcohol withdrawal. However, it has yet to be determined whether these increases in dihydropyridine binding represent changes in L-type Ca^{2+} channel protein or mRNA. It is also unknown whether there are brain regional differences in the regulation of Ca^{2+} channels.

Since different Ca^{2+} channel subunits may be expressed in different brain regions, the differential regulation of these subunits by chronic alcohol exposure may imply a differential susceptibility of certain brain regions to neurotoxicity. Chronic ethanol exposure has been found to produce tolerance to the *in vitro* inhibitory effects of ethanol on Ca^{2+} uptake. However, the development of tolerance is pronounced in some brain areas such as hypothalamus and does not occur in other areas such as cerebellum.

Interestingly, striking increases in dihydropyridine binding are seen in heart tissue of ethanol-dependent rats. This may be related to the cardiovascular effects of alcohol that have been reported in humans. There is considerable evidence that chronic heavy alcohol consumption is associated with cardiovascular disease such as cardiomyopathy, hypertension, and arrhythmia. Any or all, of these afflictions could be related to Ca^{2+} channel dysregulation caused by alcohol dependence. Since different Ca^{2+} channel subunits may be expressed in distinct tissues (for example, the heart versus the brain), the differential regulation of various L-type channel subunits may have implications for ethanol pathology that reach beyond the CNS. Cells in culture have been used in vitro models for the study of chronic ethanol effects on voltage-sensitive Ca^{2+} channels. Increases in L-type Ca^{2+} channel binding sites have been reported in both adrenal chromaffin cells and PC12 cells (a clonal line of neural crest origin) after the cells were grown in 200 mM ethanol for 6 days. It is believed that such changes in Ca^{2+} channel binding after chronic ethanol exposure constitute an adaptive response to ethanol because acute alcohol decreases voltage- dependent Ca^{2+} influx.

Although the biochemical mechanisms involved in this upregulation of dihydropyridine binding sites is as yet unclear, several studies implicate the involvement of a PKC. The increase in dihydrophyridine binding sites can be prevented by inhibitors of protein and mRNA synthesis suggesting that transcription of some gene(s) may be involved in the response. It has yet to be shown, however, that regulation of the genes encoding Ca^{2+} channel subunits, *per se*, is a component of the-adaptive process.

Dihydropyridines appear to bind to the a, subunit of the L-type Ca^{2+} channel. However, it is currently unknown how changes in dihydropyridine binding relate the changes in al or other subunit protein or mRNA. Also, since no ligands exist that specifically bind to other L-type channel subunits, it is unknown whether other subunits are changed following chronic alcohol exposure. Innate differences in susceptibility to dependence and withdrawal have been demonstrated (withdrawal seizure-prone and withdrawal seizure-resistant mouse strains exist). These innate differences reflect differences in genes and in their regulation and are important in explaining human variations in the susceptibility to alcohol dependence and withdrawal.

Differences in the effect of chronic alcohol treatment on dihydropyridine- binding sites in withdrawal seizure-prone and withdrawal seizure-resistant mouse strains have been observed implying that these innate genetic differences in the susceptibility to dependence and withdrawal may be explained, in part, by the differential regulation of voltage-sensitive Ca^{2+} channels. Thus, voltage-gated Ca^{2+} channels likely contribute to the acute actions of ethanol as well as to the development of tolerance to and dependence on ethanol. Innate differences in these channels may contribute to innate differences in sensitivity. Significantly more research will be required to understand the role of various subtypes of voltage-gated Ca^{2+} channels in ethanol responses.

Effects of ethanol on voltage-gated K^+ channels

Shefner and Osmanovic studied the effects of ethanol on locus coeruleus neuronal inward-rectifier currents and found that ethanol shifted the inward rectification in the depolarizing direction, essentially increasing current. Ethanol-induced hyperpolarization of hippocampal neurons has also been suggested to be caused by increased K^+ conductance. Similarly, single-channel currents studied using the patch-clamp cell-attached technique found that ethanol increased the open probability of K^+ channels in human T cells at concentrations of 35-50 mM ethanol.

In contrast, other researchers have found that low concentrations of ethanol selectively inhibit a voltage-dependent K^+ current known as the M current (a muscarinic-cholinergic responsive noninactivating current) in hippocampal CA1 neurons. Ethanol had little or no effect on other K^+ conductances in CA1 pyramidal cells. In this preparation, ethanol increased muscarinic-cholinergic excitation at ethanol concentrations in the 10-22 mM range by potentiating the inhibition of the M current. The M current inhibition could be due to an ethanol effect through G proteins. These studies clearly suggest that the large diversity of K^+ channels contains certain subtypes that are particularly sensitive to ethanol and that certain ones are potentiated whereas others inhibited. A low threshold transient outward K^+ current in many invertebrate and vertebrate neurons is referred to its the A current. This current appears to be important in reducing the frequency of' repetitive spike firing. A number of studies have found ethanol increases the decay time at concentrations greater than 200 mM. Interestingly ethanol appears to act directly on the channels and not through membrane effects, and different I_A currents show different sensitivity consistent with differential subunit composition mediating differential sensitivity to ethanol.

Triestman's group expressed 10 different voltage-gated K^+ channels in oocytes from the Shaker, Shab, and other families and found none of them to be particularly sensitive to ethanol, requiring concentrations of more than 200 mM to observe effects in most cases. Covarrubias and Rubin investigated the effects of ethanol on four cloned homologous K^+ channels expressed in oocytes and found that ethanol selectively inhibited only one, a noninactivating Drosophila Shaw2 channel. Ethanol rapidly and reversibly blocked this channel in a concentration-dependent manner (17-170 mM). When a series of alcohols was studied, the alcohols did not follow a linear potency hydrophobicity profile, suggesting a hydrophobic pocket within the channel.

Alcohol sensitivity could be transferred to hybrid proteins containing portions of ethanol-sensitive Shaw2 and homologous ethanol-insensitive subunits, suggesting that ethanol acts directly on the protein and not within the membrane. The direct effect of ethanol on K^+ channel subunit proteins was also suggested by studies on an ethanol-insensitive human K^+ channel clone (Kv3.4). Butanol was found to inhibit long duration single-channel openings in a manner consistent with a direct effect on the channel.

Chapter 7

Neuronal Cytoskeleton

Morphology of each cell type is defined by the organization of its intracellular components. In the past, harsh fixation methods rendered accurate determination of the structure and mechanism of these components difficult. Gray observed a reticular net that extended throughout the cytoplasm, which was described as an artifact resulting from the precipitation of a protein complex during fixation. Using high voltage electron microscopy, Porter and coworkers described a complex gellike structure that maintains the organization and intracellular movement of cell organelles, which wits named "microtrabecular lattice". Cryofixation and deep-etching techniques allowed investigators to study cytoskeletal structure in axons. Perfusing the brain with sucrose before performing the rapid freezing technique developed by Moor *et al.* allowed the cytoskeletal structure of the perikaryon and dendrites of neurons to be observed.

The cytoskeleton was found to be a complex network consisting of protein components, microfilaments, intermediate filaments, and microtubules, which not only maintain the shape of cell bodies but also regulate cytoplasmic and cellular movement. The neuronal cyctoskeleton consists of microtubules that can interact with membrane proteins including neurotransmitter receptor and ion channel. Ligand -gated ion channels, such as nictotinic acetychlorine receptors, glycine receptors glutamine receptors and γ-aminobutryic acid$_A$ (GABA$_A$) receptors, are known to cluster in plasma membranes.

Studies suggest that postsynaptic ligand-gated channels form clusters that are anchored in the plasma membrane by interacting with cytoskeletal components and these clusters may serve to optimize delivery of neurotransmitters to the channels. The interaction of clustered ligand-gated ion channels with cytoskeletal components may also play a role in channel function.

For example, studies suggest that the interaction of microtubules with GABA$_A$ receptors regulates GABA binding affinity. Regulation of neurotransmitter function may be significant in the study of neuropathological processes, such as Alzheimer's disease, neurotrauma, and experimental epilepsy, in which the cytoskeleton is vulnerable to disruption. Many types of neurotransmitter receptors are localized in clusters rather than being distributed evenly over the cell surface. This localization of receptors is controlled primarily by the cytoskeleton and likely has implication for regulation of neuronal functioning. About 25 years ago the first studies of receptor clustering demonstrated that nicotinic acetylcholine receptors are unevenly distributed and appear as hot spots throughout the plasma membrane of chick microfibers. McMahan and Wallace were the first to demonstrate that agrin, a neurally derived factor causes nicotinic acetylcholine receptors to aggregate opposite the nerve terminals at the developing neuromuscular junction.

Since that finding, other peripheral proteins have been suggested to interact with agrin to form a complex that immobilizes nicotinic acetylcholine receptor clusters at the neuromuscular unction. Glycine receptors, glutamate receptors, and γ-aminobutyric acid receptors aggregate at postsynaptic densities, and clustering of voltage gated, sodium channels occurs at the neuromuscular junction. The clustering of ligand-gated receptors opposite the nerve terminal is believed to optimize the delivery of neurotransmitters to receptors. However, the anchoring of clusters in the plasma membrane may also lay other roles in receptor function. For example, one study demonstrated hat the inter-action of the glycine receptor complex with microtubules regulates ligand binding affinity.

$GABA_A$ receptors solubilized from rat brain bind microtubules directly and are immobilized in clusters throughout the plasma membrane of spinal neurons at innervated and noninnervated sites. The purpose of this review is to discuss the role of the cytoskeleton in the clustering and function of $GABA_A$ receptors, but it is first necessary to briefly review the complex biology of the cytoskeleton. Because of the vast literature in this area, we have not attempted to be comprehensive but try to cite review articles that provide greater detail than is possible here.

Intermediate filament

Intermediate filaments intermediate between microtubules and microfilaments. These filaments form a net around the nucleus and extend out to the cell membrane that provides mechanical support to the cell body and nucleus. Intermediate filaments are categorized into four main types. Type I consists of two subfamilies of keratin-the acidic and the basic. Type II includes desmin, vimentin, and glial fibrillary acidic protein. Type V intermediate filaments are nuclear laminas consisting of sheets of filaments that assemble and disassemble only during certain stages of mitosis. Type III intermediate filaments are composed of three distinct polypeptides of different molecular masses with a diameter of 10 nm. These filaments are found in perikaron axoplasm, and dendrites, frequently in close association with microtubules in axons and dendrites, suggesting that the rigidity and strength of axons results from neurofilament-microtubule complexes.

Microfilaments

Microfilaments are a major component of the cytoskeleton in postsynaptic densities and densities and are under constant dynamic regulation, which suggests that actin dynamics may be important in regulating postsynaptic function. Microfilaments, sometimes called F actin, are polymers of a globular protein subunit, called globular actin, with a molecular mass of 42 kD. In Purkinie cells of the chick cerebellum, the cytoskeleton consists of a microfilament mesh that encircles the nucleus and other organelles and helps anchor them in the cytoplasm. In these cells the organelles located in dendrities, especially the endo-plasmic reticulum, are also attached to a mesh of microfilaments. Much more is known about the microfilament-associated proteins compared to the proteins that interact with microtubules and intermediate filaments.

Cross-linking proteins connect microfilaments in the cytoplasm of most cells. Filamin, the most abundant of these proteins is a long, flexible molecule consisting of two identical polypeptide can insioined together at the head, with a binding site for microfilaments at the tail end. In fibroblasts, different proteins such as tensin, a-actinin and talin bind microfilaments; directly and have been proposed to play a role in a complex mechanism that anchors the cells to the extracellular matrix.

Ankyrin and spectrin, microfilament-associated components of the membrane cytoskeleton, aid in immobilizing integral plasma membrane proteins and also bind voltage-gated Na channels in brain and help anchor these channels in plasma membrane. Walker et at demonstrated that the 43-kDa protrill that is associated with acetylcholine receptors binds actin, and they also proposed that this p is involved in anchoring acetylcholine recept evidence indicates that microfilaments bind directly to integral plasma membrane proteins such as the ponituculin glycoprotein of slime molds and the epidermal growth factor receptor.

Microtubules

Microtubules found in axons, dendrites, the mitotic apparatus, and the cytoplasm are considered to be labile microtubules. Labile microtubules are in a state of "dynamic equilibrium" in which the cytoplasmic pool of tubulin is in equilibrium with assembled microtubules. Microtubules are formed by polymerization of tubulin, the building block of microtubules, and can be classified into two general categories: labile and stable. A free tubulin subunit concentration provides a steady-state condition in which the addition of dimers to microtubules is balanced by the loss of dimers from other microtubules. This pool of tubulin is termed the "critical concentration."

At tubulin levels greater than the critical concentration microtubules will grow, whereas concentrations less than the critical concentration will cause microtubules to shrink. Furthermore, labile microtubules are rapidly depolymerized by micro tubule-disrupting agents such as colchicine and by low temperatures.

In contrast, stable microtubules are not in a state of dynamic equilibrium and are not affected by low temperaturc or microtubule depolymerizing agents.Tubulin, the building block of microtubules, consists of two globular monomer subunits, a and b each with a molecular mass of 55 kDa, which bind covalently to form a dimer. Tubulin polymerizes to form protofilaments, with each microtubule consisting of 13 protofilaments aligned in parallel to form a hollow tube. Labile microtubules are always in a state of dynamic instability as a result of microtubule assembly and disassembly.

Each tubulin molecule in the soluble pool is bound to two guanosine triphosphates (GTPs), one of which is stable and the other, readily hydrolyzed. The polymerization of microtubules depends on the delayed hydrolysis of GTP and not on the energy provided by GTP hydrolysis. At the growing end of microtubules, tubulin containing GTP, which promotes microtubule assembly, has a higher affinity for tubulin than does tubulin containing guanosine diphosphate (GDP). During the elongation of microtubules at the growing end, tubulin polymerizes faster than GTP can hydrolyze. Once the GTP is hydrolyzed at the growing end microtubules begin to shrink. This mechanism allows individual microtubules to exhibit abrupt phases of elongation and shortening. Microtubules have a defined polarity, with each end being functionally and structurally different.

Polymerization and depolymerization of microtubules occur more rapidly at the (+) end than at the (-) end. In vitro, tubulin, will be added twice as fast to the (+) end than to the (-) end. However, in the cell cytoplasm during interphase, microtubule polymerization occurs only at the distal (+) end because the proximal (-) end is contained within the microtubule-organizing center, which lies just outside the nucleus and harbours the centriole. As the cell enters mitosis, the microtubule network rearranges, forming a mitotic spindle in which the (-) ends of the microtubules are connected to one of the microtubule-organizing centers, called the mitotic poles. In axons of neurons, the (-) end of microtubules begins at the centriole near the axon hillock. However, microtubules that originate within the axon are not connected to a centriole, yet maintain the same polarity.

Microtubules contain a hetero-geneous class of proteins known as microtubule-associated proteins (MAPs), which promote tubulin Polymerization and stabilize microtubules. More than 20 different MAPs have been identified in the brain and the specific localization of MAPs to axons and dendrites has been proposed to play an important role in defining cell morphology.

The most abundant of the MAPs are MAP-2 and tau, which are located predominately in dendrites and axons, respectively. MAP-2 not only stabilizes microtubules but also cross-links them to other microtubules, neurofilaments, and intermediate filaments. This cross-linking function of MAP-2 resides in two functional domains, the microtubule binding region, which binds directly to 12 tubulin dimers, and the projection arm, which extends away from microtubules and probably cross-links microtubules to other component of the cytoskeleton.

Matus demonstrate that transfecting epithelial cells with MAP-2 altered the structure of these nonneuronal microtubules, which developed characteristics of microtubules in dendrites—they were stiffer and longer than the microtubules in nontransfected cells. MAP-2 regulates tubulin polymerization by a phosphorylation-dependent mechanism. Phosphorylation of MAP-2 in vivo by cAMP dependent protein kinase or Ca^{2+}/ calmodulin dependent kinase diminishes the ability of MAP-2 to promote tubulin assembly as demonstrated by reconstitution studies. The affinity of MAP-2 for microtubules is also decreased by phosphorylation of MAP-2 by cAMP-dependent kinase or protein kinase C.

The phosphorylation of brain tubulin by Ca^{2+}/ calmodulin dependent kinase inhibits the ability of tubulin to bind MAP-2 and to polymerize into microtubules. In addition, tubulin can also be phosphorylated by protein kinase A. Tau increases the rate of tubulin polymerization and stabilizes microtubules with projections that can cross-link adjacent microtubules, thereby allowing microtubules to organize into bundles that are necessary for the formation of axons. Transfecting cultured fibroblast cells with tau caused microtubules to align into bundles, similar to those in nerve cells. The interaction of tau with microtubules is also regulated by phosphorylation.

Phosphorylation of tau by cAMP-dependent kinase and Ca^{2+}/calmodulin depen-dent protein kinase diminishes its ability to promote microtubule polymerization. Abnormally phosphorylated tau proteins are the major component of the paired helical filament, which are the principal fibrous structures of the neurofibrillary lesions that are characteristic of Alzheimer's disease. These neurons contains fewer microtubules and more pathological tau. The abnormal tau proteins can be distinguished from the normal form by the presence of the many phosphorylated sites.

In contrast to normal tau proteins, abnormally phosphorylated tau proteins are phosphorylated at Ser 396, which decreases the binding affinity of tau for microtubules and its ability to stabilize microtubules Bramblett. This abnormal form of tau has been found in fetal brains, which suggest that in Alzheimer's disease the phosphorylation of tau may be caused by inappropriate activation of fetal kinases or inactivation of tau protein phosphatases.

Biernat *et al.* demonstrated that phosphorylation of tau proteins at Ser 262 by 35/41 kD kinases also strongly reduces binding affinity of tau protein for microtubules. Furthermore, this form of tau reacted with paired helical filament antibodies, which suggest that this site is abnormally phosphorylated in Alzheimer's disease. Using biopsy-derived adult brain tau, Matsuo showed that tau from paired helical filaments is phosphorylated at almost all of the same sites as is normal tau from adult brain.

However, in autopsy-derived adult brain, normal tau is phosphorylated to a much lesser extent than is abnormal tau, which indicates that certain protein phosphatases are downgraded in Alzheimer's disease patients. This suggests that the production of paired helical filament tau is the result of the decreased efficiency of phosphatases that dephosphorylate tau, which leads to the accumulation of the hyperphosphorylated tau. Yet, this study found that only one of the many paired helical filament antibodies recognized a site on the paired helical filament tau but not on the normal tau.

The epitope recognized by this antibody and the phosphorylation site within the epitope still remain to be identified. Gephyrin, a 93-kD protein associated with the glycine receptor, binds polymerized tubulin. Froehner proposed a model for the anchoring of glycine receptors in which gephyrin links glycine receptors to the underlying microtubules. In spinal neurons and kidney 293 cells co-transfected with gephyrin and glycine receptors, gephyrin is required for the immobilization of glycine receptors into aggregates in the plasma membrane. Gephyrin, also, is located in many areas of the brain, and by northern blot hybridization, gephyrin transcripts have been revealed in brain, liver, kidney, and lung; furthermore, five different splice variants of gephyrin mRNA were identified in brain.

These data suggest that gephyrin may also link other types of integral plasma membrane proteins to microtubules not only in brain but also in other tissue types. In similarity with microfilaments, microtubules can interact directly with integral plasma membrane proteins. Item and Sieghart demonstrated that rat brain $GABA_A$ receptors bind microtubules directly and suggested that microtubules are responsible for anchoring $GABA_A$ receptors in plasma membrane.

Function of microtubules

During mitosis, formation of the mitotic spindle and radiation of microtubules from the two centrosomes are essential for the proper separation of the chromosomes, replicated during S phase, to the daughter cells and for cell division. During interphase, microtubules radiate throughout the cytoplasm from the microtubule-organizing center, which contains the centriole. These microtubules not only aid in anchoring organelles in cytoplasm but also act as cables along which mitochondria and smaller membrane bound organelles change their position.

In the cytoplasm of cultured monkey kidney epithelial cells the interaction of microtubules with the endoplasmic reticulum is responsible for the reticular formation. In chick embryo fibroblast cells the endoplasmic reticulum vesicles are aggregated near the nucleus immediately after mitosis. During interphase, microtubules are repolymerized, and endoplasmic reticulum vesicles slide along the stationary microtubules to form reticular branches by interacting with intersecting microtubules and fusing with other branches of endoplasmic reticulum.

Axons contain thick bundles of microtubules produced by the cross linking of tau proteins that extend throughout the axoplasm. All the microtubules have the same polarity, with the (-) end extending from the centriole near the cell body. In neurons, ribosomes are present only in the cell body and dendrites, and, thus, proteins are not synthesized in axons but are transported down the axon to the synaptic terminal by axonal transport. Transport to the synaptic terminal is in the anterograde direction [(-) end to (+) end of microtubules] and involves growth of axons and replacement of synaptic vesicles. In the opposite retrograde direction [(+) end to the (-) end], synaptic vesicles and proteins, destined for degradation, are transported to lysosomes in the cell body.

The slowest moving proteins are cytoskeletal proteins such as tubulin and inter mediate-filament proteins, which are transported at a rate of approximately 0.2-1 mm/day. Actin, clathrin, calmodulin, and myosin move down axons; at a rate of 2-50 mm/day. Fast axonal transport consists of vesicles, smooth endoplasmic reticulum, and small granules moving at a rate of 200-400 mm/day. Fast axonal transport involves the interaction of microtubules with organelles and the motor proteins, kinesin or MAPIC. These motor proteins act as cross-bridges between microtubules and vesicles, and in the presence of ATP, the kinesin proteins move along the microtubule in the anterograde direction transporting the bound vesicle.

Dendrites consist of bundles of microtubules linked by MAP-2 proteins to other microtubules, microfilaments, or intermediate filaments. However, unlike axons, dendrites consist of microtubules with non-uniform polarity and also contain the organelles necessary for protein synthesis such as the endoplasmic reticulum and Golgi apparatus. This suggests that microtubules aid in the distribution and motility of organelles in the cytoplasm of dendrites.

Microtubules are present in presynaptic terminals and form coils that interact closely with mitochondria and the plasma membrane, which suggests that the microtubules play a role in translocating synaptic vesicles to the active zone. However, earlier studies demonstrated that, in intact neurons, microtubules terminate short of the synaptic terminals. This discrepancy may result from the type of fixation techniques used in these different studies. In the earlier studies, neuronal preparations were routinely fixed and dehydrated at extremely low temperatures and in solutions containing high concentrations of Ca^{2+}. The exposure of microtubules to low temperatures and high Ca^{2+}, which easily penetrates aldehyde-treated membranes, leads to microtubule depolymerization. In contrast, Hajos *et al.* prevented microtubule depolymerization by fixing synaptosomes at room temperature, whereas Gordon-Weeks *et al.* preserved microtubule coiled formations by using albumin, a Ca^{2+} chelator.

Immunocytochemical assays revealed high concentrations of actin in postsynaptic densities and dendritic spines and MAP-2 in dendritic spines. Similar techniques demonstrated that tubulin exists in postsynaptic densities, which are located on the internal surface of the postsynaptic membrane and likely play a role in immobilizing integral plasma membrane proteins. Furthermore, microtubules have also been shown to associate with membranes. These plasma membrane-associated microtubules may aid in the anchoring of ligand-gated ion channels.

Cytoskeleton-affecting agents

Colchicine

It is an alkaloid obtained from the plant *Colchicium autumnale* (autumn crosus, meadow saffron) and various plants of the Liliaceae family. Preparations of *Colchicium* were reported to be used to relieve articular pain as early as the sixth century A.D. However, because of its toxicity, it was not until the eighteenth century that Colchicium was used for the treatment of acute gout. It continues to be widely used for this purpose today.

In 1889, Pernice discovered that Colchicium was an antimitotic agent and during the 1930s, colchicine was shown to arrest cellular division in metaphase, which led to its widespread use for the study of mitosis. Taylor and his collaborators illustrated that the tubulin dimer is not only the building block of microtubules but also the binding site of colchicine and that the antimitotic action of colchicine is derived from its ability to disrupt microtubules. Colchicine depolymerizes microtubules by binding to tubulin, thereby decreasing the soluble tubulin pool that is in equilibrium with the assembled microtubules and inhibiting microtubule polymerization. Each tubulin dimer contains at least two high-affinity binding sites for colchicine on the 3 subunit. Furthermore, at 37°C the colchicinetubulin complex is formed at a maximal rate and prevents the polymerization of additional tubulin.

The binding of colchicine is slowly reversible, leading to the depolymerization of labile microtubules. The more stable a microtubule is, the less its structure is affected by colchicine. The effects of colchicine on microtubule depoly-merization are temperature and concentration dependent. At 0°C, colchicine does not depolymerize microtubules for two reasons: (a) Colchicine does not bind tubulin at 0°C and (b) Labile microtubules are already completely depolymerized at this temperature. At 37°C, colchicine binds to tubulin and depolymerizes microtubules at a maximal rate. Furthermore, a short incubation with a high concentration of colchicine (1001μM) both inhibits the ability of disassembled microtubules to repolymerize and depolymerizes assembled microtubules. However, low concentrations of colchicine (1-101μM) can inhibit the repolymerization of microtubules but are not capable of depolymerizing assembled microtubules after short incubation times.

Colchicine consists of three rings. The C ring is required for tubulin binding activity, whereas the A and B rings are not required for the interaction of colchicine with tubulin but do play a role in the affinity of colchicine for tubulin because modifications in these rings decrease affinity for tubulin.

Modifications of the C ring to other types of aromatic rings inhibit the activity of the compounds. Irradiation with ultraviolet light converts colchicine to β- and γ- lumicolchicines, which do not bind to tubulin and thus do not inhibit the polymerization of microtubules. Lumicolchicines have been used in a number experiments as control agents to test the specificity of colchicine on microtubules.

Vinblastine

Like colchicine, vinblastine is used as an antimitotic agent because it inhibits the ability of microtubules to polymerize. Vinblastine is a vinca alkaloid that is extracted from the periwinkle plant (Vinca rosea) and used as a chemo-therapeutic agent. Incubation with vinblastine results in the depolymerization of microtubules and the formation of highly regular vinblastine-tubulin crystals that consist of one mole of vinblastine per mole of tubulin. The binding site of vinblastine on tubulin differs from that observed with colchicine.

Nocodazole

Nocodazole (methyl [5-(2-thienylcarbbnyl)-1 H-benzimidazol-2-yl] carbamate, R17934) is a synthetic drug with antimitotic activity. Nocodazole rapidly enters cells and binds to free tubulin leading to depolymerization of microtubules and inhibition of polymerization. Binding studies indicate that nocodazole competitively inhibits colchicine binding to purified brain tubulin even though these agents are not structurally similar.

Taxol

Taxol was originally isolated from the stem bark of the western yew Taxus brevifolia and has antileukemic and tumor inhibitory properties. Taxol, like colchicine, is an antimitotic agent. However, instead of depolymerizing microtubules, taxol promotes microtubule assembly and protects them from depolymerization by cold, colchicine, remain polymerized under conditions in which they usually disassemble, such as the absence of GTP, the presence of high Ca^{2+}, and low temperatures.

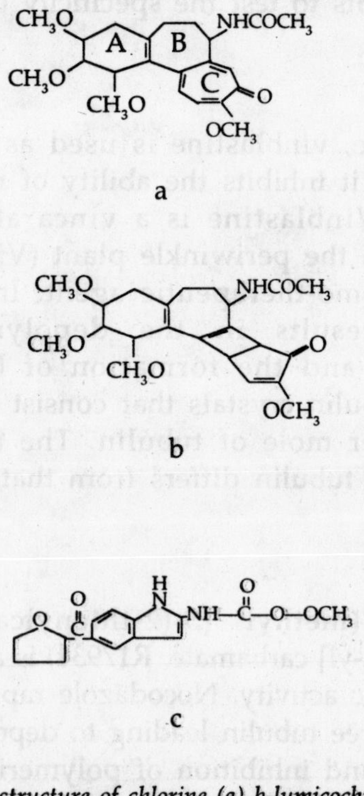

Fig 1: Chemical structure of chlorine (a) b-lumicochicine (b)vinblastine (c) Ring designations for colchicine

Cytochalasins

Cytochalasins (cytochalasin B and cytochalasin D) are members of a family of i metabolites, excreted by various molds, that bind specifically to the fast-growing end of microfilaments and prevent the addition of actin. These agents induce the rundown of *N*-methyl-D-aspartate (NMDA) responses of cultured. Phalloidin is a poisonous alkaloid produced by the toadstool Amanita phalloides. In contrast to the cytoplasms, phalloidin stabilizes microfilaments, thereby inhibiting depolymerization. Also, unlike cytochalasins, phalloidin is not membrane permeable and thus must be injected intracellularly. Phalloidin was found to stabilize NMDA channel rundown.

Cytoskeleton and GABA$_A$ receptors

GABA$_A$ receptors have also been shown to cluster and to be immobilized in rat spinal neurons and plasma membranes of rat cerebellar granule cells. GABA$_A$ receptors consisting of bovine $\alpha_1\beta_1\gamma_{2L}$, bovine $\alpha_1\beta_1$ human $\alpha_1\beta_2$ human $\alpha_1\beta_2\gamma_{2L}$, human $\alpha_1\beta_2\gamma_{2s}$, or human $\alpha_1\beta_2\gamma_2$ subunits formed clusters throughout the plasma membrane of stably transfected mouse L(tk$^-$) cells. This is a significant finding because the clustering was not clearly apparent for various GABA receptor constructs transfected in kidney 293 cells. Microtubules were found to prevent the lateral mobility of clustered GABA$_A$ receptor clusters and anchor GABA receptor clusters in plasma membrane. A key question is whether microtubules immobilize clustered GABA$_A$ receptors in the membrane by a direct interaction of microtubules with clustered GABA$_A$ receptors or via a linker protein such as gephyrin.

Gephyrin colocalizes with clustered GABA$_A$ receptors in spinal neurons, sympathetic preganglionic neurons, and in retina. However, gephyrin does not bind to GABA receptor subunits isolated from brain and is not sufficient for aggregating cotransfected GABA receptors in kidney 293 cells. Furthermore, Item and Sieghart demonstrated that GABA receptors isolated from rat brain bind micro-tubules directly. These data suggest that microtubules immobilize GABA receptor clusters by direct interaction with GABA receptors. $GABA_A$ receptors form clusters on the surface of neuronal membranes and are associated with tubulin. This raises the question of whether disruption of the microtubule cytoskeleton. All alter the function of these receptors.

Recent studies demonstrated that the microtubule-disrupting agents tested (colchicine, viriblastine, taxol, and nocodazole) inhibited muscimol-stimulated chloride uptake by brain microsacs and reduced the function of various $GABA_A$ receptor constructs expressed in Xenapus oocytes or in stably transfected cells. Colchicine, vinblastine, nocadozole, and taxol, which have different chemical structures and disrupt micro-tubules through different mechanisms, all inhibited the function of $GABA_A$ responses. In contrast, the lumi-colchicines, α- or γ-lumicolchicine, did not inhibit GABA responses. Colchicine did not have an effect on $GABA_A$ receptors reconstituted in proteoliposomes, a system that does not contain intact microtubules.

To determine which pharmacological characteristic of the GABA site was regulated by microtubules, several studies examined the effects of microtubule depolymerization on both $GABA_A$ receptor function (muscimol EQo) and $GABA_A$ receptor affinity (using the $GABA_A$ receptor antagonist SR 95531) in brain microsacs and stably transfected cells. These studies showed that depolymerization of microtubules in cortical microsacs or stably transfected cells decreased both the potency of muscimol to stimulate chloride uptake and the affinity of SR 95531 binding by approximately 50%, without decreasing either the efficacy of muscimol at the $GABA_A$ receptor or the density (B_{max}) of SR95531 binding sites. Another question is whether microtubule depolymerization could alter the action of other pharmacological sites of the $GABA_A$ receptor. For example, benzodiazepines and barbiturates aıe allosteric modulators of $GABA_A$ receptor function and it was of interest to determine whether the actions of drugs on these sites are dependent upon intact nucrotubules. In one study, microtubule depolymerization did not alter the potentiating action of either flunitrazepam or pentobarbital in stably transfected cells and did not alter the binding parameters of flunitrazepam or TBPS in cortical microsacs.

However, the depolymerization of microtubules did inhibit ethanol sensitivity in stably transfected cells. Because the site of ethanol action is not yet known, it is difficult to determine the mechanism by which microtubules regulate ethanol action. Thus, the GABA-binding site is the only pharmacologically defined domain of the $GABA_A$ complex found thus far to be regulated by microtubules.

There are at least two possible mechanisms by which disruption of the cytoskeleton can alter the affinity of GABA for its receptor. (a) Protein-protein interactions between receptor complexes alter the binding affinity, which would suggest that it is the clustering that is critical. (b) The interaction between microtubules and the receptor complex affects the affinity which would imply that clustering is not important but is an epiphenomenon that merely reflects the integrity of the cytoskeleton.

One study suggests that the second possibility is correct, namely, that microtbules regulate both the lateral mobility of clustered $GABA_A$ receptors and the GABA affinity of these receptors, but these are separate phenomena. Clustered $GABA_A$ receptors were cross-linked with antibodies to examine whether it is the lateral interaction between $GABA_A$ receptors within clusters or the interaction between microtubules and $GABA_A$ receptors that allosterically modulates the GABA binding site. Based on the data from the cross-linking experiments, Whatley suggested that the direct interaction of microtubules with $GABA_A$ receptors, independent of receptor clustering, produces a long-range allosteric modulation of the GABA binding site that increases GABA binding affinity.

Neurapatholgy and cyloskeleton

Variety of neuropathologic conditions can affect the cytoskeleton. Central excitotoxicity, resulting from release of the amino acid neurotransmitter glutamate, produces neuronal deterioration and death in many pathological conditions. Hypoxia/ischemia or hypoglycemia impair energy metabolism and decrease the ability of the neuronal and glial uptake systems to remove synaptically released glutamate rapidly from the extracellular space. Brain trauma or infection is characterized by axonal injuries that leak excitatory amino acids into the extracellular space.

Neuropathological disorders have been shown to produce epileptic seizures in the limbic system, which further increases the concentration of extracellular glutamate. This toxic exposure of postsynaptic dendrites to glutamate can lead to cell swelling and eventually neuronal degeneration through the activation of the glutamate receptors NMDA, kainate, and AMPA. These receptors are permeable to potassium and sodium, and furthermore, the NMDA and kainate channels have a high calcium permeability. In cultured hippocampal cells, intense glutamate exposure produces immediate neroural swelling, that can be prevented by the removal of extracellular sodium and chloride.

However, even without the acute neuronal swelling, the majority of the cells exposed to excessive glutamate are destroyed. Removal of extracellular calcium substantially decreases the glutamate-induced neuronal loss. Thus, these data suggest that the initial swelling is due to excessive influx of sodium through the glutamate channels, which leads to an increased entry of chloride and water, and that the neuronal degenerative process is mediated by an excessive influx of calcium through the NMDA channels.

Studies demonstrated that exposure of neuronal cells to glutamate and the NMDA receptor antagonists D-2-aminophosphonovalerate (APV) or ketamine did not prevent the initial neuronal swelling but markedly inhibited neuronal cell loss. However, compounds known to antagonize quis-qualate or kainate receptors did not reduce glutamate neurotoxicity, supporting the idea that excessive influx of calcium is responsible for neurodegeneration.

Intra-cellular calcium is the central regulator of neuronal cytoskeletal structure. High levels of calcium cause a breakdown in the cytoskeleton, which leads to disruption of the transport of organelles and the structure of the plasma membrane. Microtubule and microfilament polymerization is regulated by intracellular calcium.

Neither microtubules nor microfilaments will polymerize at a low level of intracellular calcium (50 n*M* in cultured hippocampal neurons) but will do so at a moderate concentration of intracellular calcium (70-120 n*M*). However, at higher calcium concentrations (200-250 n*M*) the outgrowth of both microtubules and microfilaments ceases. Very high calcium levels (>300 n*M*) result in neuronal degeneration, characterized by disruption of the cytoskeleton, and eventually cell death. Alzheimer's disease is characterized by the loss of basal forebrain cholinergic neurons that project to the cerebral cortex, hippocampus, and amygdala. Injection of glutamate into the basal forebrain of rats results in pathomorphological changes similar to those observed in Alzhrimer's disease. Furthermore, 14 months after injection of glutamate, paired helical filaments appeared in various limbic and neocortical regions.

Exposure of hippocampal cells to high glutamate or a calcium ionophore (A23187) resulted in elevation of intracellular calcium, reduction in microtubules, increased in reactivity of tau with paired helical filament antibodies, and decreased in reactivity of tau for normal tau antibodies. Colchicine produced alterations in the antigenic properties of tau similar to those induced by calcium. After calcium-induced alterations in tau, phosphatases or taxol decreased the immunoreactivity of tau toward the paired helical filament antibodies and increased the immunoreactivity toward normal tau antibodies. However, phosphatases did not reverse the immunoreactivity of tau induced by colchicine to the immunoreactivity of normal tau. These results indicate that excessive intracellular calcium or the depolymerization of microtubules independent of intracellular calcium concentrations produced antigenic changes in tau similar to those in the abnormal tau of paired helical filaments.

Stabilizing microtubules with taxol prevented high intracellular calcium levels from producing antigenic alterations in normal tau. Thus, the depolymerization of microtubules by calcium may produce the paired helical filaments present in Alzheimer's disease. The cytoarchitectural disturbances of the limbic structures of patients with schizophrenia are the result of brain damage during cortical development.

Chapter 8

Neuronal Plasticity

The growing neurites possess, at their distal tips, a structure called the growth cone, which consists of highly motile filipodial protrusions that extend from a central base. The structure and motility of the growth cone are controlled by the underlying cytoskeleton, which consists of actin filaments concentrated in filopodia and lamellipodia. Data indicate that the actin based filopodial system is involved in the directionalization of outgrowth and neurite branching. In contrast to the growth cone itself, which lacks microtubules, the neurite shaft is replete with microtubules and contains few actin filaments. The polymerization of microtubules in the neurite shaft is the primary mechanism for elongation of the neurite. Synapses are formed when an axonal growth cone encounters a postsynaptic cell and reciprocal signalling mechanisms are engaged that locally inhibit further outgrowth and promote differentiation of the synapse.

Free radicals and neuronal plasticity

Ffree radicals and nonradical ROS play important roles in regulating neurite outgrowth, synaptogenesis, and synaptic plasticity. First, ROS have been shown to affect a variety of regulatory systems involved in the control of neurite outgrowth and synaptic plasticity. Second, environmental signals believed to play important roles in regulating growth cone behaviours and synaptic function (e.g., neurotransmitters and neurotrophic factors) can induce or suppress the production of ROS in neurons. Third, antioxidants and other agents that affect ROS production or metabolism can affect neurite outgrowth, synapto-genesis, and synaptic transmission, in various systems. Cytoskeletal components are sensitive to modification by ROS.

For example, microtubules and intermediate filaments were reported to be disrupted in cultured rat cardiac myocytes following exposure to 4-hydroxynonenal. Neurotransmitter release can also be regulated by ROS. For example, nitric oxide and hydroxyl radical modulate acetylcholine release from cultured cortical neurons. Cell differentiation is highly sensitive to levels of ROS. For example, St. Clair. *et al.* showed that cultured mouse embryo fibroblasts over-expressing Mn-SOD exhibit greatly enhanced differentiation into muscle like cells.. The roles of ROS in the differentiation of neurons are largely unknown.

Pro or antioxidants can affect neurite outgrowth and synaptic plasticity. Smalheiser el al. reported that phospholipase A, and lipoxygenase inhibitors prevented acute neurite retraction in response to lysophosphatidic acid and several other agents in cultured neuroblastoma cells; the inhibitors also decreased F-actin levels. Exogenous arachidonic acid promoted neurite retraction. Arachidonic acid is generated in cells under conditions that induce transmitter, release.

In synaptosomes prepared from rat cerebral cortex, arachidonic acid induced a marked increase in phosphorylation of the growth-associated in the GAP-43. Arachidonic acid syner with calcium in promoting protein kinase C-mediated phosphorylation of GAP-43. As phosphorylation of GAP-43 has been stimulation of axonal outgrowth.

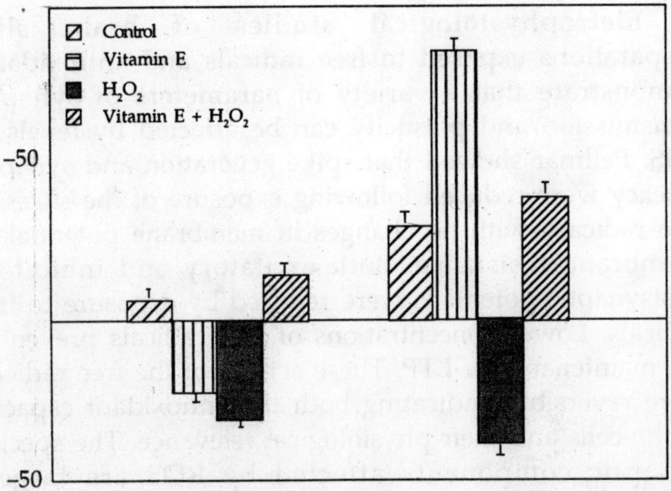

Fig. 1. Differential modulation of dendrites and axonal outgrowth by reactive oxygen species

Nitric oxide is a free radical implicated in the regulation of synaptic plasticity, and data also suggest a role for NO in the regulation of neurite outgrowth. Hess. *et al* reported that nitric oxide rapidly inhibited in a reversible manner, neurite outgrowth in cultured dorsal root ganglion neurons. In the latter study it was also shown that NO inhibits activation of neuronal proteins, suggesting a possible post-translational mechanism for the rapid modulation of growth cone proteins involved in the regulation of neurite outgrowth. An additional mechanism whereby free radicals might affect neurite outgrowth, synaptogenesis, and cell survival is the inhibition of enzymes and enzyme inhibitors. For example, Bolkemus and Monard reported that xanthine oxidase-derived free radicals inhibited protease nexinI and that antioxidants counteracted this effect. Protease nexin I is a neurite outgrowth-promoting, astrocyte derived inhibitor of the enzyme trypsin that has been shown to modulate calcium homeostatic and neuronal vulnerability to excitotoxic and oxidation.

Eletrophysiological studies of brain slice preparations exposed to free radicals and antioxidants demonstrate that a variety of parameters of synaptic transmission and plasticity can be affected by levels of ROS. Pellmar showed that spike generation and synaptic efficacy were reduced following exposure of the slices to free radicals, with no changes in membrane potential or membrane resistance. Both excitatory and inhibitory postsynaptic potentials were reduced by exposure to free radicals. Lower concentrations of free radicals prevented the maintenance of LTP. These actions of the free radicals were reversible, indicating both the antioxidant capacity of the cells and their physiological relevance. The specific synaptic components affected by ROS are largely unknown. However, redox modulation of the NMDA receptor likely plays at least a partial role.

The antioxidant a-tocopherol (vitamin E) induced a slowly developing, long-lasting increase of the EPSP in guinea pig hippocampal slices. NMDA receptor antagonists did not block the induction of LTP by vitamin E, suggesting a mechanism independent of NMDA receptors. Nitric oxide has rapidly become the most well documented free radical regulator of synaptic plasticity, wherein it may act as a retrograde messenger that is generated in the postsynaptic cell and then diffuses into the presynaptic terminal where it modulates neuro-transmitter release. In studies involving intracellular recordings from individual pre- and postsynaptic neurons in hippocampal cell cultures, Arancio *et al* provided evidence that NO induces UP by directly acting on presynaptic terminals. In this study, application of the NO scavenger oxymyoglobin to the culture medium or inside either the presynaptic or postsynaptic cell blocked LTP. Injection of a NOS inhibitor (N -nionomethyl-t-arginine) into the postsynaptic cell blocked LTP, whereas injection of the same NOS inhibitor into the presynaptic cell had no effect on LTP.

Experiments, in which a photo releasable form of NO was injected into the postsynaptic neuron and then released immediately prior to tetanic stimulation showed that postsynaptic production of NO results in an enhancement of the excitatory postsynaptic potential, indicating enhanced neurotransmitter release from the presynaptic terminal. Cyclic GMP may mediate the effect of NO in the presynaptic cell because injection of cGMP into the presynaptic cell produced activity-dependent potentiation of synaptic transmission. Studies of mice rendered genetically deficient in NOS have provided further evidence of a role for NO in LTP. Whereas mice lacking only neuronal NOS (nNOS) exhibited no Impairment in UP, mice lacking both nNOS and endothelial NOS (eNOS) exhibited a clear reduction in LTP, although it did still occur. ROS may activate more conventional signaling pathways that then secondarily modulate developmental and synaptic plasticity.

cGMP mediates some effects of NO on neurite outgrowth and synaptic plasticity. Following the exposure of cultured embryonic hippocanipal neurons to sodium nitroprusside, there is an increase in cGMP levels as determined by radioimm-unoassay of cell homogenates or immuno-cytochemistry. The cCGMP immunoreactivity is present throughout neurons and is particularly concentrated in growth cones. Elevation of cGMP levels appears to play a role in the modulation of dendrite outgrowth by sAAPa and glutamate. Whole-cell perforated patch-clamp recordings from cultured embryonic hippocampal neurons showed that cGMP activates high-conductance, charbydotoxin-sensitive K^+ channels. In addition, cGMP suppresses NMDA induced currents in cultured hippocampal neurons.

Studies of the responses of CAI hippocampal neurons to stimulation of Schaffer collateral inputs showed that LTP can be blocked by an inhibitor of soluble guanylate cyclase. Zhuo *et al* found that an inhibitor of soluble guanylate cyclase does not block the NMDA component of the exicitatory postsynaptic potential, but appears to affect LTP by a presynaptic mechanism. Recordings of field potentials from the dendrites of CAI neurons in response to stimulation of Schaffer collateral axons indicate that cCMP can modulate LTD. Thus, whereas a 10-Hz stimulation normally does not induce LTD, it does induce LTD in slices pretreated with 8-broinocGMP.

Interestingly, previous data indicated that sAPPa does not elevate cGMP levels by activating soluble guanylate cyclase; rather, sAPPa activates a membrane-associated guanylate cyclase. It is therefore conceivable that cGMP generated in response to SAPPa may act locally within the immediate vicinity of the plasma-membrane to regulate ion channels such as the NMDA receptor and, K^+ channels.

Role of $NF_K B$ in neuronal plasticity

$NF_K B$ appears to be in a unique position to mediate long-term, spatially defined changes in neuronal plasticity. The reason is that, in contrast to other trascription factors that are normally localized in the nucleus. This means that $NF_K B$ is located at sites such as growth cones and-postsynaptic dendrites, where processes mediating developmental and synaptic plasticity are initiated. $NF_K B$ is present in the cytoplasm of neurons and glial cells in the brain in both constitutively active and inducible forms. In embryonic rat hippocampal and cortical cell cultures, there is a high constitutive activity of $NF_K B$ in neurons that is correlated with the presence of p65 and p50 in nuclei of the neurons. Similarly, neurons throughout rat and mouse brain exhibit readily detectable basal levels of activated $NF_{\kappa}B$; interestingly, there is considerable heterogeneity in basal levels of activated $NF_{\kappa}B$ among neuronal populations, suggesting a relationship to cell function.

Further evidence for the constitutive activation of $NF_{\kappa}B$ in brain cells in vivo comes from studies showing high levels of human immunodeficiency virus type I promoter in several brain regions of transgenic mice. Kaltschmidt and co-workers have shown that inducible forms of $NF_{\kappa}B$ (p65—IKB complexes) are present in neurons, wherein a considerable portion of the NFKB is localized to synapses. Thus, synaptosomal preparations from rat cerebral cortex contained consider able amounts of inducible $NF_{\kappa}B$, and double-labelling immuno-histochemical analyses showed colocalization of $NF_{\kappa}B$ subunits, and the synaptic vesicle-associated protein synaptophysin.

$NF_{\kappa}B$ can be activated by excitatory neuro-transmitters, membrane depolarization, and synaptic activity. Exposure of cultured cerebellar granule neurons to the ionotropic glutamate receptor agonist kainate or KCl (depolarization) resulted in a marked activation of $NF_{\kappa}B$, as determined at the single-cell level by using an antibody that selectively reacts with the activated nuclear form of p65.

Interestingly, the latter study also showed that activation of $NF_{\kappa}B$ could be blocked by an antioxidant, indicating that free radicals were involved in glutamate receptor-and depolarization-induced activation of $NF_{\kappa}B$ obtained similar findings and in cerebellar gramile neurons as well. Whether $NF_{\kappa}B$ is activated at glutamatergic synapses during normal activity throughout the brain remains to be determined.

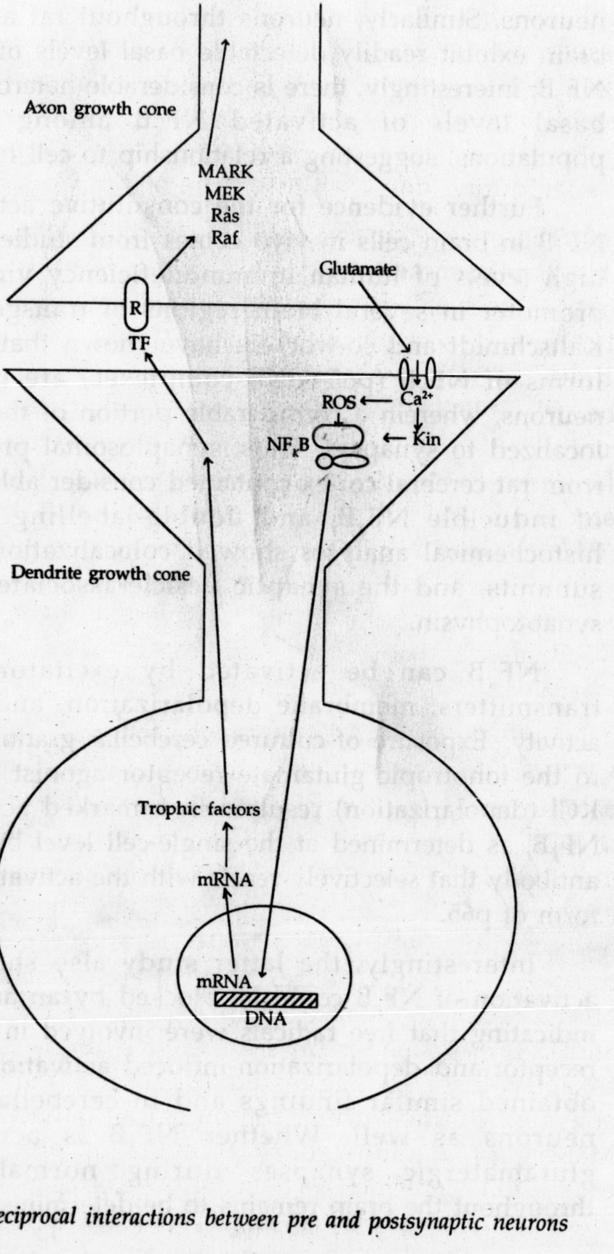

Fig. 2. Reciprocal interactions between pre and postsynaptic neurons

Roles for $NF_\kappa B$ in the development of many different tissues are suggested by studies establishing a major role for the Drosophila $NF_\kappa B$ homologue "dorsal" in the establishment of dorsoventral polarity in the developing embryo. Levels of $NF_\kappa B$ activity have been found to change during development of the nervotis system.

The peak $NF_\kappa B$ activity occurred during the period of synaptogenesis, suggesting a role for $NF_\kappa B$ in that process. Inducible 050 dimers and p65—cRel dimers were shown to be present in the brains of young rats, but not adults. Schmidt *et al.* generated transgenic mice in which the if lacZ gene, as driven by the p105 promoter, which contains an $NF_\kappa B$ site lacZ expression was first seen in the cerebellar nuclei and spinal medulla, followed by later expression in the pontine nuclei. In the adult mice, evidence for $NF_\kappa B$ activation was present in the hippocamptis and some cortical regions. Although not yet examined, the data are also consistent with a role for $NF_\kappa B$ in programmed developmental neuronal death, which occurs during the period of synaptogenesis.

Indeed, studies have shown that the activation of $NF_\kappa B$ can prevent neuronal death in various cell culture paradigms. In the mature nervous system, $NF_\kappa B$ appears to play an important role in synaptic plasticity, as indicated by studies of $NF_\kappa B$ activation in paradigms of hippocampal LTP. The induction of $NF_\kappa B$ in cortex and hippocamptis by seizure activity in adult rats indicated that high levels of neural activity are associated with $NF_\kappa B$ activation. Because synaptic activity during seizures and LTP may share many features, the ability of seizures to activate $NF_\kappa B$ suggested that $NF_\kappa B$ may also be activated during normal physiological activity. Indeed, Meberg *et al.* showed that levels of p65 and p50 expression were increased in hippocampal granule cells after LTP. However, the expression of p65 and p50 was also increased by low-frequency stimulation, suggesting that $NF_{K}B$ activation is sensitive to synaptic activation in general, rather than to LTP induction parameters in particular.

Role of calcium in neurite outgrowth

The key roles of calcium in regulating neurite outgrowth and synaptic plasticity have been studied intensively and are fairly well-understood. On the other hand, the involvement of ROS in developmental and synaptic plasticity has only recently begun to be studied and is largely uncharacterized. What is particularly intriguing is that neurons have harnessed the multifunctional properties of both calcium and ROS and utilize them in many different signalling pathways, both in the developing and in the mature nervous system. Equally intriguing is the fact that calcium and ROS act its messengers of synaptic degeneration and cell death.Intracellular free calcium plays an important role in shaping neuronal form and function from the earliest stages of cell differentiation.

Early events in the morphological development of neurons are the sprouting of neurites and their differentiation into axons and dendrites. Neurite sprouting can be induced or sup-pressed by calcium influx, depending upon the intensity and duration of the influx stimulus.

Manipulations that prevent calcium influx or cause excessive calcium influx can prevent neurite sprouting in cultured invertebrate and vertebrate neurons. The role of calcium in regulating the establishment of neuronal polarity (i.e., differentiation of dendrites and the axon) has been studied in cultured embryonic hippocampal pyramidal-like neurons. Hippocampal neurons elaborate several short processes that are morphologically indistingui-shable; subsequently (after 24-48 hr, in culture) one process will accelerate its outgrowth rate and differ-entiate into the axon. A gradient of intracellular calcium concentrations appears to play an important role in determining which process differentiates into the axon for the following reasons:

i) the axon has a lower resting Ca^{2+}, than the other processes;
ii) exposure of cells to agents that disrupt the calcium gradient prevents axon formation; and
iii) experimentally induced focal elevations in [Ca^{2+}] induce the axon to form at sites most distant from the site of calcium influx.

During a critical period of neuronal development, the property of cell polarity remains maleable and subject to modulation by environmental factors. Embryonic hippocampal neurons in culture that have established polarity can be induced to change their polarity by altering the calcium gradient in the cell. Transections of the initial axon- with a micropipette will cause one of the shorter processes (presumptive dendrites) to differentiate into a new axon. Calcium influx at the site of transection of the axon appears to play a major role in inducing morphological "repolarization" because it sets up a new calcium gradient within the cell that is highly correlated with the site of new axon formation (the new axon forms where [Ca^{2+} is lowest) and because the prevention of calcium influx at the site of transection of the axon prevents a new axon from forming.

Environmental signals that influence the establishment of neuronal polarity, in vivo remain to be identified. However, it appears that ongoing electrical activity and activation of glutamate receptors are not required for the establishment of polarity. Very early during the process of neuronal differ-entiation the cells become excitable and responsive to neurotransmitters, neurons also produce and release neurotransmitters in a calcium-dependent manner during the very early stages of neurite outgrowth and well prior to synaptogenesis.

Studies of mechanisms regulating neurite outgrowth in cultured invertebrate and vertebrate neurons identified calcium as a key mediator of growth cone responses to neurotransmitters and electrical activity. When cultured *Helisoma* (pond snail) neurons were exposed to the excitatory neurotransmitter serotonin calcium influx occurred and growth cone, advance was inhibited; manipulations that prevented Ca^{2+} influx blocked the ability of serotonin to inhibit neurite growth.

Similarly, electrical depolarization of Helisoma neurons in culture suppressed neurite outgrowth in a Ca^{2+} dependent manner. Further studies showed that filipodial activity (actin system) and neurite elongation. (tubulin system) are differentially sensitive to Ca^{2+}. For example, Ca^{2+} influx induces rapid polymer-ization of actin (filipodial extension), but at the same time inhibits tubulin polymerization (suppression of neurite elongation). This dual action Ca^{2+} may account for the response of growth cone to cones in their environment that signal synaptogenesis.

Subsequent studies of embryonic mammalian-neurons in culture showed that calcium is a key mediator of outgrowth responses to the excitatory neurotransmitter glutamate. Glutamate selectively inhibited dendrite outgrowth, without affecting axon outgrowth, in cultured embryonic rat hippocampal pyramidal-like neurons. The reason that dendrite outgrowth is selectively affected by glutamate is that the glutamate receptors linked to, calcium influx are concentrated, in the dendrites and are not present in the axon. Stimuli that induce the release of calcium from intracellular stores can also induce neurite retraction. Further experiments, in which interactions between axons from entorhinal cortex neurons and target hippocampal neurons were studied, showed that glutamate released from the axons acted to inhibit dendrite outgrowth and promote synaptogenesis.

Studies of developing visual pathways have confirmed the important role of glutamate in the regulation of synaptogenesis and have further shown that the NMDA receptor plays a particularly important role in developmental-synaptic plasticity. Neurotrophic factors that promote neurite outgrowth may also act by modulating intracellular calcium levels. For example, basic fibroblast growth factor (bFGF) induces the outgrowth and branching of cultured embryonic rat hippocampal neurons. Long-term (many hours) pretreatment of cultured hippocampal neurons with bFGF prior to exposure to glutamate also counteracts the dendrite outgrowth-inhibiting action of glutamate by a mechanism linked to the suppression of calcium responses to glutamate.

Other neurotrophic factors that may also modulate growth cone behaviours by calcium-mediated mechanism include nerve growth factor (NGF), brain-derived neurotrophic factos (BDNF), and insulin like growth factors. The localization of glutamate receptors within neurons appears to play an important role in determining the sites where synapses can for in and which sites ultimately differentiate into "stable" synapses.

Developing embryonic hippocampal pyramidal neurons will segregate NMDA receptors to dendritic and somatic compartments (receptors are excluded from the axon.); this compartmentalization of NMDA receptors occurs in single isolated neurons, indicating that ell-cell interactions are not required. Interactions between neurons result in further compartmentalization of NMDA receptors such that they localize to presumptive postsynaptic sites along dendrites; this segregation process likely involves the release of endogenous glutamate.

The physical localization of glutamate receptors is tightly associated with sites of activity-dependent calcium influx as shown by Bekkers and Stevens. As might be expected from consideration of the intricacy of the neuritic arbors and synaptic connections that define the unique cellular architecture of the nervous system, regulation of growth cone behaviours and synaptogensis is tightly controlled, both spatially and temporally.

Davenport and co-workers showed that individual filipodia isolated from their parent growth cone can respond to membrane depolarization, demonstrating the presence of voltage-dependent Ca^{2+} channels in individual filipodia. Moreover, the isolated filipodia exhibited calcium responses to dopamine and serotonin, which were attenuated by coapplication of acetylclioline. Because axonal growth cones can release neurotransmitter locally, the data suggest the possibility of highly localized neurotransmitter-mediated outgrowth-regulating mechanisms in growth cones.

Studies of the role of calcium in synaptic plasticity have been performed in a variety of systems, ranging from cell culture and ex vivo studies of invertebrate neurons to electrophysiological analyses of neural circuits in adult mammals. Information that appears to be particularly relevant to the role of calcium in regulating the synaptic correlates of learning and memory in humans has come from studies of hippocampal slices from adult rodent brains. Hippocampal circuitry is such that the axons of delitate granule neurons (mossy fibers) synapse on distal dendrites of CA3 pyramidal neurons; the axons of CA3 neurons in turn form synapses on distal dendrites of CA1 pyramidal neurons. Each of these synapses employs the excitatory amino acid neurotransmiitter glutathate.

Collingridge and co-workers recorded postsynaptic response in dendritic fields of CA1 neurons before and after stimulation of the Schaffer collaterals and found that stimulation at relatively high frequencies (e.g., 30-50 Hz) resulted in enhanced postsynaptic response to subsequent stimulation. This phenomenon, called long-term potentiation (LTP), has since been linked to learning and memory, and the molecular underpinnings of LTP have been studied intensively.

It is clear that UP requires both calcium-mediated release of glutamate from the presynaptic terminal and calcium influx little the postsynaptic neuron. Calcium influx through the N methyl-D-aspartate (NMDA) receptor plays a key role in the induction of LTP. The postsynaptic calcium influx leads to the activation of kinases such as calcium/calmodulin depended protein kinase C (PKC), which have been mechanistically linked to the induction of LTP.

The substrates phosphorylated by these kinases that are critical in inducing enhanced synaptic effectiveness have not been established. Stimulation of axonal input to CA1 hippocampal neurons is particularly effective in inducing UP because of a strong postsynaptic response resisting from the temporal summation of dendritic excitatory post synaptic potentials; a threshold level of activation of NMDA receptors is required to trigger LTP. Stimulation of the Schaffer collaterals at frequencies below the threshold required to induce UP (1-5 Hz) induces long-term depression (LTD) of the active synapses.

As with LTP, NMDA receptor activation and calcium influx into the postsynaptic neuron are required for the induction of LTD. When the stimulation frequency is varied and the total number of pulses is held constant, it is possible to establish a relationship between stimulation frequency and synaptic plasticity. At low frequencies (1-5 Hz) LTD is induced, at a frequencies of approximately 10 Hz neither LTD or UP is induced, and at high frequencies (greater than 20 Hz) LTP is induced.

On the basis of studies that employed inhibitors of protein kinases and phosphatases, it has been proposed that large increases in postsynaptic calcium induce UP by activating protein kinases and that small increases in postsynaptic calcium induce LTD by activating, protein phosphatases.

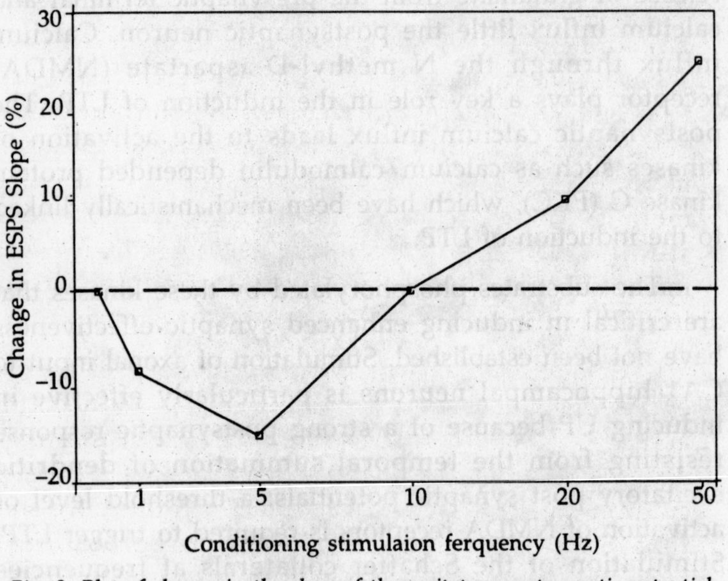

Fig. 3. Plot of change in the slope of the excitatory postsynaptic potential as a function conditioning stimulus frequency

Many calcium-mediated signalling pathways that regulate growth cone behaviours and synaptic plasticity involve post-translational modification of proteins involved in rapid changes in neuronal form and function. For example, calcium can activate CAMKII, which then phosphorylates substrates, such as the microtubule-associated proteins and MAP-2 and actin-binding proteins in the regulation of neurite elongation and growth cone motility. Calcium mediated phosphorylation of ion channels modulates neuronal excitability and thereby influences the responsiveness of growth cones to environmental signals that affect channel function.

Activation of postsynaptic CaMKII plays an important role in the induction-established by studies in which CaMKII of UP in the hippocampus, as inhibitors blocked LTP. Moreover, a deficiency in mice rendered genetically induction of LTP occurs.

Another calcium-sensitive enzyme that has been linked to both regulation of neurite outgrowth and synaptic Plasticity is PKC. Phorboi ester activators of PKC and PKC inhibitors can either inhibit induce neurite outgrowth, depending upon the cell type. Roles for PKC in bip-pocanipal TP and other paradigms of learning and memory are also suggested by electrophysiological studies that employed pharma-cological or genetic manipulations of PKC or that examined PKC activation in LTP paradigms.

Calcium can also induce the activation of adenylate cyclase, resulting in cyclic AMP (cAMP) production; this action of calcium may serve to CAMP responses to agonists linked to the GTP-binding protein, Gs. Studies of cultured invertebrate and vertebrate neurons have shown that CAMP regulates growth cone behaviours. In Helisoma neurons, CAMP inhibits neurite outgrowth by a mechanism involving the stimulation of calcium influx through plasma membrane channels.

In embryonic rat hippocanipal pyramidal neurons, CAMP stimulates dendrite outgrowth and antagonizes the- dendrite outgrowth-inhibiting action of glutamate. At mature synapses, CAMP appears to play a prominent role in the events underlying learning and memory.

Early studies in Aplysia demonstrated a central role for CAMP in associated learning at specific synapses, and subsequent studies of hippocanipal slices from adult rats provided evidence of an important role for CAMP in the induction of LTP. Activity-dependent regulation of neurite outgrowth is not simply the result of rapid local actions of calcium at sites where it enters cells. Rather, the activity and calcium influx trigger an array of signalling pathways that mediate long-term responses of the neurons.

Calcium stimulates several different signalling cascades that lead to the activation of transcription factors. For example, calcium activates adenylate cyclase via a calmodulin-mediated pathway, leading to the generation of CAMP. Cyclic AMP then activates CAMP-dependent protein kinase (PKA), which activates a CAMP response element-binding protein (CREB) that induces the transcription of various cAMP-responsive genes.

Calcium may also activate kinases such as CaMKII within the nucleus, resulting in the activation of transcription factors such as Elk-1 SRF (serum response factor), and CREB. Another example of a nuclear signalling pathway activated by calcium involves Ras; Ras activates the serine-threonine kinase, Raf Raf initiates a cascade of phosphorylation reactions involving MEK-I and mitogen-activated protein (MAP) kinase, a pathway commonly used by growth factors that bind receptor tyrosine kinases.

Thus, calcium is a critical regulator of synaptic plasticity as it controls neurotransmitter release from the presynaptic terminal, as well as both short- and long-term postsynaptic responses involved in processes such as UP and LTD.

Synaptic degeneration

Cell degeneration in many, if not all, paradigms of neuronal cell death in the mammalian CNS is initiated at excitatory synapses. This hypothesis is based on the simple fact that the synapse is the site where calcium influx and kOS production are concentrated during normal neuronal activity. Activation of NMDA and AMPA-kainate receptors at synapses results in large increases in $[Ca^{2+}]$ (10- 100 μM) that are highly localized to the environment of the postsynaptic spine.

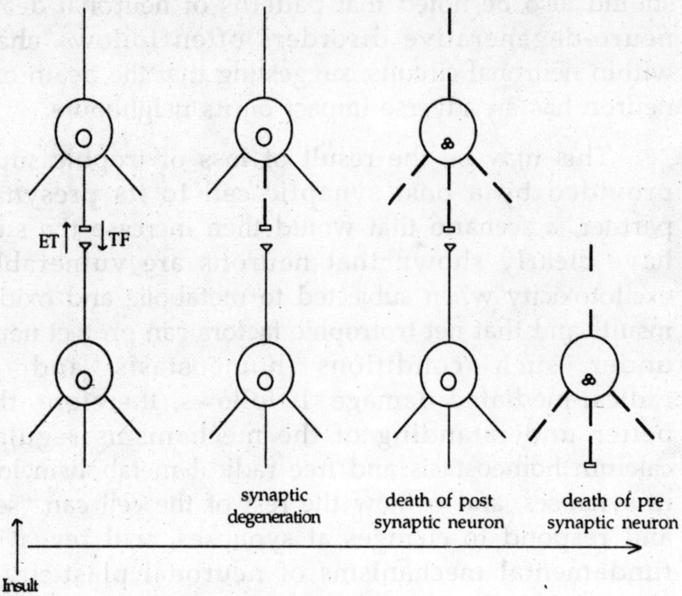

Fig. 4. Synaptic degeneration

Mitochondria, which are major sources of ROS, are present in the postsynaptic sites and are known to generate O_2^- and H_2O_2 in response to calcium influx induced by glutamate receptor activation. Blockade of synaptic activity by glutamate receptor antagonists protects hippocampal neurons against a variety of calcium and ROS-mediated insults, including exposure to ischemic and hypoglycemic conditions, mitochondrial toxins, iron, and amyloid strongly suggest a major contribution of synaptic activation of glutamate receptors to cell death in these paradigms.

Studies of postmortem brain tissue from humans with neuro-degenerative conditions, ranging from severe temporal lobe epilepsy to stroke to, Alzheimer disease, have documented severe dendritic degeneration with axonal sparing, consistent with postsynaptic sites being the regions where the degenerative process begins. It should also be noted that patterns of neuronal death in neuro-degenerative disorders often follow "chains" within neuronal circuits, suggesting that the death of one neuron has an adverse impact on its neighbours.

This may be the result of loss of trophic support provided by a post synaptic cell to its presynaptic partner, a scenario that would then increase the studies have clearly shown that neurons are vulnerable to excitotoxicity when subjected to metabolic and oxidative insults and that net trotrophic factors can protect neurons under such conditions homeostasis and free, radical-mediated damage. It follows, therefore, that a better understanding of the mechanisms regulating calcium homeostasis and free radical metabolism locally in synapses, and of how the rest of the cell can "sense" and respond to changes at synapses, will reveal both fundamental mechanisms of neuronal plasticity and potential molecular targets for preventing neuronal degeneration in lathophysiological settings.

Calcium influx plays a central role in excitotoxic neuronal death, a degenerative process initiated at synapses. A remarkably broad array of insults results in elevationsm [Ca^{2+}] that contribute to the degeneration and cell death caused by those insults. Perhaps the best-documented paradigm is glutamate toxicity, wherein calcium imaging studies have shown that glutamate causes a large and sustained deviation in [Ca^{2+}] that precedes cell degeneration and death. Glutamate-induced calcium influx through plasma membrane ionotropic glutamate receptors and VDCC is causally involved in the cell death process because the removal of extracellular calcium and compounds that; block glutamate receptor channels and VDCC attenuate glutamate toxicity.

The events downstream of calcium that promote cell degeneration and death include activation of proteases and generation of ROS. Metabolic insults reveal to the pathogenesis of ischemic neuronal degeneration (stroke), such as glucose deprivation, hypoxia, and mitochondrial toxins, also induce increased [Ca^{2+}] which contributes to neuronal degeneration. As described earlier, [Ca^{2+}), can rise to 10-100 gm locally if postsynaptic sites. Moreover, ROS accumulation appears to reach much higher levels in postsynaptic regions than in the cell body or axons of cultured hippocampal neurons exposed to metabolic insults. Under physiological conditions, levels of calcium and rapidly recover following a period of synaptic activity.

The recovery is due to the presence of efficient mechanisms for the removal of calcium and ROS in the post-synaptic region. For example, immunocytochemical studies have shown that levels of certain calcium-binding proteins and ion-motive ATPases are greater in synaptic regions, than elsewhere in the cell. Whether antioxidant systems are also enriched at synaptic sites has not been established, but seems likely. Indeed, a comparison of antioxidant enzyme levels in cortical synaptosomes and total cortical homogenates suggests an enrichment of catalase and a lower level of Cn/Zn-SOD in synapses. The localization of antioxidant enzymes within synapses remains to be determined. Under pathophysiological conditions, however, systems for the removal of calcium and ROS ultimately fail, resulting in sustained local elevations in calcium and ROS that can then spread to the rest of the cell. A particularly important point to bear in mind is the reciprocal relationship between calcium and ROS such that elevations in $[Ca^{2+}]$ promote ROS production and, conversely, ROS disrupt ion homeostasis. Calcium promotes the accumulation of O_2, H_2O_2, OR, NO, and peroxynitrite by the pathways. ROS, in turn, disrupt ion homeostasis in several different ways. For example, ROS can impair the function of the plasma membrane Na^+/K^+ and Ca^{2+} -ATPases in cultured embryonic rat hippocampal neurons, which leads to membrane depolarization and the inability to restore resting $[Ca^{2+}]$ Oxidative stress can also impair the function of the neuronal glucose transporter GLUT3. Which results in ATP depletion, secondary failure of ion-motive ATPases, and increased neuronal vulnerability to excitotoxicity.

An additional system compromised by ROS is glutamate transport, which promotes excitotoxic cascades by inducing the accumulation of extracellullar glutamate. Excessive elevations in $[Ca^{2+}]$ can kill neurons by either apoptosis or necrosis, depending upon the severity of the magnitude and duration of the $[Ca^{2+}]$ elevation. Exposure of cultured hippocampal or cortical neurons to high concentrations of excitatory amino acids induces rapid and sustained elevations in $[Ca^{2+}]$ that are associated with necrosis a form of cell death characterized by cell swelling, organellar degeneration, and cell. On the other hand, exposure of the same neurons to much lower levels of glutamate or to metabolic and oxidative insults that promote excitotoxic cascades results in more subtle elevation in [Ca^{2+}] and delayed apoptotic death. Apoptosis is characterized by cell shrinkage, nuclear condensation, and DNA fragmentation, with maintenance of the integrity of the plasma membrane and organelles. Many studies of non neuronal cells have implicated calcium in apopiotic paradigms.

In several different neuronal apoptosis paradigms calcium also appears to play an important role. Trophic factor withdrawal induced apoptosis may involve elevations in [Ca^{2+}], as suggested by data showing that calcium-binding proteins and inhibitors proteases prevent the NGF withdrawal-induced death of cultured sympathetic neurons. Data from studies of natural cell death of motor neurons during the development of the spinal cord also implicate calcium in apoptosis. Amyloid β-peptide induces apoptosis in cultured hippocampal and cortical neurons by a mechanism that involves calcium influx through plasma membrane channels and calcium release from intracellular stores. Calcium release from irresponsive endoplasmic reticulum stores appears to play a role in several apo-ptotic paradigms.

As evidence, dantrolene protected cultured hippocampal neurons against apoptosis induced by amyloid peptide. Moreover, expression of a mutant form of the presenilin I protein in cultured PC12 cells resulted in enhanced calcium release from IP3-sensitive stores and increased vulnerabilty of neurons to apoptosis induced by trophic factor withdrawal and amyloid β-peptide. The HIV coat protein gp120 induced neuronal apoptotis in hippocampal cell cultures that was mediated by NMDA receptor activation and calcium influx. As evidence, calpain inhibitors and the calcium binding protein calbindin protect cultured cells against apoptosis induced by trophic factor withdrawal. NGF caused a decrease in calpain activity in cultured PC12 cells.

Kalin *et al.* reported that levels of mRNA encoding the IP: recieptor and IP: receptor protein were increased in B and T lymphocytes induced to undergo apoptosis in response to anti-Igm antibodies. ThIp3 receptor apparently plays an important role in the apoptotic process because the treatment of T lymphocytes with IP3 receptor antisense oligodeoxynucleotides prevented dexamethasone-induced apoptosis. Perhaps not too surprisingly, calcium influx itself activates signaling pathways designed to stabilize calcium homeostasis and promote neuronal survival. It has been known for some time that elevation of extracellular K^+ levels, a condition that depolarizes the plasma membrane, can promote long-term neuronal survival in cell culture.

Depolarization can also prevent cell death induced by trophic factor withdrawal and amyloid β-peptide, and the mechanism apparently involves increased calcium influx, which induces the expression of neuroprotective gene products. Indeed, transient calcium influx may play a role in the cytoprotective actions of some neurotrophic factors. Many genes involved in regulating neuronal plasticity and survival contain response elements sensitive to calcium-activated transcription factors such as AP-1 and CREB, and some of these surely play a role in the activity-dependent regulation of neuronal survival. Additional evidence for the importance of calcium in synaptically driven apoptotic and necrotic neuronal deaths comes from studies that established relationships between the expression of calcium-binding proteins and resistance to cell death and from complementary data showing that manipulations that suppress elevations in $[Ca^{2+}]$ can prevent glutamate-mediated neuronal degeneration.

Chapter 9

Neurobiology of Stroke

Over the past several years, experimental studies have provided new and important data regarding patterns of neuronal vulnerability following cerebral ischemic insults. These data have significantly contributed to our understanding of the pathogenesis of ischemic, neuronal injury and provided new insights regarding treatment strategies. From the perspective of a neuroscientist, experi-mental investigations have uncovered the complexity of the response of the central nervous system to acute injury and the multifactorial nature of the injury process. This chapter will summarize the clinical features of stroke and provide a discussion of the current status of stroke therapy.

The definition of cerebral infarction includes stroke due to cerebral embolism and thrombosis. Atherothr-ombotic brain infarction, defined as, infarction, secondary to large-vessel atherothrombosis as well as lacunar infarction, comprises the most common stroke subtype. Intraparenchymal and subarachnoid haemorrhages not associated with structural lesions such as intracranial neoplasm or vascular malformations are other common types of stroke. International studies of stroke mortality demonstrate a steady decline in the death rate due to stroke. Between 1970 and 1985, stroke mortality rates declined in most countries for both men and women, but were usually greater in women.

Since 1979, however, the annual rate of decline of stroke mortality has slowed considerably in the United States. Stroke incidence increases with age and is generally higher among men than women. In the Framingham study, the incidence of all strokes and atherothrombotic brain infarctions specifically was about 30% higher in men than in women. Although stroke mortality increases with age, it is especially high among certain racial and ethnic groups, such as Afro-Americans and Asians, and may be higher among men except in the oldest age group.

Stroke risk factors may be subclassified into several categories, including single risk factors, multiple risk factors; well-documented risk factors, and treatable and nontreatable risk factors. Most information concerning risk factors comes from the Framingham study, and an excellent review of these risk factors has been published. Atherothrombotic brain infarctions (ABI) comprise the most common stroke subtype, accounting for 44% of strokes.

Studies have emphasized the importance of the degree of asymptomatic carotid stenosis as a determinant of stroke risk. Thus, with carotid stenosis of less than 75%, the stroke rate is low (1.3%), whereas with carotid stenosis of greater than 75%, the stroke rate increase to 3.3% annually. International studies have shown that hypertension, whether systolic, disastolic, or combined systematic and systolic, is an important risk factor for ischemic stroke. In the Framingham study, a history of hypertension was a powerful predictor of brain infarction in men an women.

Thus, the incidence of stroke, and to a lesser extent atherothrombosis, is increased with isolated systolic hypertension. Heart disease, including coronary heart disease, congestive heart failure, left ventricular hypertrophy, and arrhythmias such as atrial fibrillation, is also considered an important risk factor for stroke. Diabetes mellitus is risk factor for ischemic stroke of the large-artery type, but is of question able significance in other stroke subtypes. For atherothrom-botic stroke, the, impact of diabetes mellitus is greater for women than for men. In terms of race, Asians of Japanese and Chinese descent and Afro-Americans have higher stroke rates than white Americans.

Although the reasons for the ethnic disparity in stroke rates are not known, differences may be environmentally rather than genetically determined. The relationship between blood lipids and cerebral infarction is also controversial. In some segments of Western society, there is evidence linking hypercholesterolemia and increased low-density lipoproteins to ischemic stroke. However, this is not true for all study populations. A relationship between a high normal hematocrit level and increased incidence of brain infarction was also suggested by the Framingham study. The relationship holds only for men 35-64 years of age.

Other studies have linked serum fibrinogen level to the incidence of stroke. A potential relationship between apolipoprotein E (APOE) genotype and dementia in patients with stroke has been described. Whereas some studies have reported a higher frequency of APOE than controls, other investigations have found no association between APOE genotype and stroke.

Overall, patients with dementia and stroke have a higher APOE4 frequency than controls, implying that individuals who develop dementia with stroke share genetic susceptibility. In addition, APOE4-related risk of dementia and stroke was different in Whites than in other ethnic groups, in parallel with findings from Alzheimer disease patients.

Future studies are needed to clarify the role of vascular risk factors and disease in relation to APOE genotype. In terms of alcohol consumption, a J-shaped curve has been proposed to describe the association between moderate alcohol consumption and the risk of ischemic stroke. Previous studies have indicated that there may be an excess risk of stroke at high levels of alcohol consumption, with protective effects seen at moderate levels of consumption. Cigarette smoking is another well-described risk factor for, ischemic stroke

Mechanisms underlying the effects of smoking on stroke incidence have been summarized previously importantly, there may be substantial reduction in stroke risk with the cessation of cigarette smoking. Oral contraceptives increased the risk of ischemic stroke and that most strokes occurred in women older than 35 years with other risk factors, such as smoking and hypertension.

Prevention of stroke by the treatment of modifiable atherogenic host and environmental risk factors is likely to be the most effect means of reducing stroke morbidity and mortality. Individuals in the population with a high number of risk factors can be identified and usually treated with strategies and drugs to achieve substantial reductions in risk. In a mass approach, lifestyle modification is used to achieve modest reductions in the level of risk factors in every segment of the population. Thus, the high-risk approach can be achieved by a massive and expensive screening process that may still fail to prevent at least 50% of strokes. The mass approach is advanced through health education and economic measures to discourage exposure to various risk factors.

Because heart disease is the most important precursor of stroke, the prevention and treatment of coronary heart disease is anticipated to reduce stroke incidence. Taken in aggregate, two major randomized primary prevention trials have shown a statistically significant reduction in nonfatal myocardial infarctions and a 33% reduction in the risk of stroke, but some increase in disabling strokes in those subjects treated with aspirin. A new antiplatelet agent, ticlopidine, may be more effective than aspirin and shows substantial benefits for women. Two large-scale control trials have shown that carotid endarterectomy is the treatment of choice when there is high-grade (77-99%) ipsilateral symptomatic carotid stenosis in patients with transient ischemic attacks (TIA) or mild stroke (North American Symptomatic Carotid Endarterectomy Trial Collaborators).

Cerebral ischemia and stroke

Models have been developed to investigate the pathophysiology and treatment of cerebral ischemia and stroke. It should be stressed that although no one model adequately mimics the clinical conditions of stroke, experimental models have become very useful in investigating the pathophysiology of brain ischemia. In addition, animal models are used for the testing of novel therapeutic agents that may have potential clinical use.

Thus, although stroke rarely occurs in humans without underlying cardiovascular hematologic disease, the ability to produce ischemic pathology in relatively healthy animals has allowed investigators to investigate mechanisms of ischemic brain injury as well as target specific pathological processes for pharmacological intervention. Various factors, including animal size, interspecies variability, and anesthesia are all important factors that need to be considered when evaluating ischemia models. Many of these points have been summarized previously in helpful review articles.

Large animals are more expensive to obtain and maintain than smaller rodent species, and therefore larger animals, including nonhuman primates, dogs, and cats, are-used less frequently than rodents. Pathological outcomes in rodent models appear to be more reproducible than some in models using larger animals. Postinjury care and the ability to maintain animals for chronic survival periods are other important factors when discussing which particular species is most appropriate for a study. The metabolic rate of the brain at the moment of the ischemic insult is known to contribute to outcome, and the use of anesthetics in ischemia models is an important consideration. Because humans do not fill victim to stroke under general anaesthesia, some animal models have been developed in conscious animals.

However, most ischemic research today is done, on animals that are anesthetized and monitored physiologically. To produce global or focal ischemia in healthy animals, investigators commonly occlude one or multiple arteries by mechanical ligation or by the introduction of foreign matter into the lumen of arteries to cause embolism. Although model's incorporating mechanical occlusion of vessels have been shown to induce reproducible patterns of global or focal ischemia, these models do not incorporate thrombotic processes in the pathophysiology of the injury process. Thus, thrombotic models have also been developed that incorporate acute endothelial injury, platelet activation, and thromboembolic processes that more closely mimic the clinical condition.

Global ischemia

Models of transient global ischemia resulting in patterns of selective neuronal vulnerability are models that attempt to mimic the pathophysiology of cardiac arrest or hemodynamic conditions that result from systemic hypotension. In the two-vessel occlusion (2-VO) model of forebrain ischemia, reversible high-grade forebrain ischemia is produced by bilateral common carotid artery (CCA) occlusions combined with systemic hypotension sufficient to reduce forebrain blood flow to severe ischemic levels. The CCAs are included and blood pressures are reduced to 50 mm Hg by controlled arterial hemorrhages. Autoradiographic studies for the determination of local cerebral blood flow (lCBF) have reported reductions in blood flow to less than 5% of control in the cerebral cortex, less than 15% of control in the caudoputamen and hippocampus, and to a lesser and more variable degree in the thalamus and midbrain.

During postischemic recirculation initiated by the removal of the carotid ligatures, an immediate hyperemic phase followed by prolonged postischemic hypoperfusion has been documented in various laboratories. Variable degrees of ischemia, ranging from 10 to 30 mm, have been shown to give rise to ischemic cell changes within selectively vulnerable brain structures, including the CA1 pyramidal neurons of the hippocampus, cerebral cortex, and dorsolateral striatum. This model of transient global ischemia has the advantages of a one-stage surgical preparation and the production of a high-grade forebrain ischemia. In addition, chronic survival studies are easily conducted, thereby providing a model for assessing the effects of potential neuroprotective drugs on histopathological and behavioural outcomes. Similar to the 2-VO model, the four-vessel occlusion, (4-VO) model results in high-grade forebrain ischemia, but can be produced in awake, freely moving rats. The model is produced in two stages where a cautery needle is used to electrocoagulate the vertebral arteries, and arterial clasps are placed around each CCA and exteriorized through a ventral midline neck incision.

Forebrain ischemia can then be produced 24 hr later in awake animals that are handheld or restrained while carotid clasps are tightened. The 4-VO model has also been used in anesthetized rats. Under these conditions, the bilateral vertebral artery coagulation is conducted on the first day, and 1 day later the CCA are isolated in the anesthetized animal and transiently occluded. Like the 2-VO model, histopathological outcome involves consistent neuronal injury within the CA1 hippocampus, dorsolateral striatum, and neocortex.

Blood flow studies have shown flow markedly reduced within the striatum and neocortex to less than 3% of control and reduced to 3-7% of control in the hippocampus. Blood flow in the brain stem is maintained at approximately 25-30% of control, a result that allows for chronic survival. Similar to the 2-VO model is the finding that reperfusion is associated with early hyperemia followed by postischemic hypoperfusion that lasts for approximately 1 hr in the hiopocampus and striatum, but up to 6 hr in the parietal neocortex. Although forebrain ischemia models are of value for studying cerebral ischemia, these models do not exactly mimic the hemodynamic consequences of cardiac arrest, which results in complete ischemia to the brain, spinal cord, and extracerebral organs. Safar and colleagues initiated the development of a model of cardiac arrest in rats. Asphyxial cardiac arrest from apnea in rats yields fairly controllable insults.

However, with standard external cardiopulmonary resuscitation (CPR) and intravenous epinephrine, fewer than 30% survive at 24 hr. The total cessation of blood circulation to all organs in models of cardiac arrest is an important characteristic with regard to human cardiac cases. A reproducible outcome model of cardiac arrest with apneic asphyxia of 8 min, leading to the cessation of circulation at 3-4 min of apnea and resulting in cardiac arrest for 4-5 min. At 72 hr after injury, widespread patterns of ischemic neurons were found in many brain regions, including the cerebral cortex, caudate putamen, CA1 and CA3 subsectors of the hippocampus, thalamus, cerebellum, and brain stem. Although this asphyxia model has several limitations, the model provides a method for studying the mechanisms of neuronal injury resulting from the clinically realistic cerebral insult and screening potential cerebral resuscitation therapies.

Kawai *et al.*, introduced another model of global or cerebral ischemia associated with cardiac arrest. This model allows compression of the main cardiac blood vessels, resulting in cessation of cerebral circulation. Light microscopic neuronal changes studied in rats subjected to 10 min of complete ischemia demonstrated early neuronal changes, predominantly in GABAergic neurons. After 7 days, the hippocampus revealed the loss of approximately 50% of CA1 pyramidal neutrons. Local cerebral blood flow values generally indicate reactive hyperemia with delayed hypoperfusion seen in most structures 30 min after recirculation.

Focal ischemia

Models of focal ischemia typically giving rise to localized brain infarction have routinely been used to investigate the pathophysiology of stroke. For example, models of middle cerebral artery (MCA) occlusion in a variety of species have gained increased acceptance due to their relevance to the human clinical setting. The subtemporal approach initially investigated by Tamura and colleagues has emerged as a standard method of proximal MCA occlusion. In models of permanent MCA occlusion, electrocauterization of the MCA proximal to the origin of the lateral lenticulostriate arteries is utilized routinely.

In these models, severe reductions in ICBF are seen within the ischemic core, with milder reductions in blood flow within the border or penumbral regions. The addition of moderate arterial hypotension hits the effect of enlarging infarct volume. Models of transient MCA occlusion have also been developed whereby surgical clips or sutures are introduced to induce a transient, ischemic insult.

In human ischemic stroke, recirculation frequently occurs after focal ischemia. Thus, various models of transient focal ischemia have been developed. In rats, mechanical clipping of the MCA can be used as a model of transient occlusion. However, this technique involves invasive procedures, including craniotomy. A relatively noninvasive method of achieving reversible MCA occlusion by using an intraluminal suture has been established. In this procedure, a nylon filament suture is inserted through the proximal external carotid artery and into the internal carotid artery. After variable periods of occlusion, the suture can then be removed to promote recirculation.

Although the suture model has gained increasing popularity, infarct areas produced by this technique have been reported to vary in size and distribution, thereby necessitating the use of large numbers of animals to discern statistical significance in drug testing. A modification has-been adopted whereby the suture is coated with poly (L-lysine), a polytetonic polymerized amino acid that appears to increase adhesive forces and yield large infarcts with reduced interanimal variability. Hypertension is an important risk factor in stroke.

Although spontaneously hypertensive rats (SHR) normally are not prone to spontaneous stroke, they a re more susceptible to cerebrovascular occlusion procedures than might be tolerated in normotensive strains. Thus,

MCA occlusion gives rise to larger and more consistent cortical infarcts in SHR than in normotensive strains. In contrast, stroke-prone strains of SHRs produce cerebral hemorrhages and infarcts spontaneously in greater than 80% of the males. In these rat strains, sites of histopathological damage-include the anteromedial and occipital cortices and basal ganglia, areas representing arterial border zones. The involvement of vascular boundary zones (i.e., watershed areas) is of obvious clinical importance.

The clinical condition of stroke is commonly associated with thrombotic processes that initiate the occlusive ischemic episocle. Thus, it would be advantageous for animal models to include a thrombotic component. Photochemically induced cerebral thrombosis is an experimental approach by which, following the administration of a photosensitizing dye, local enclothelial injury with subsequent platelet aggregation can be induced by irradiating the vessel with light of a specific wavelength.

Dye-light interactions induce enclothelial damage that leads to various pathological processes, including increased blood-brain barrier (BBB) permeability and platelet recruitment and activation. Using the photochemical approach, Watson and colleague first demonstrated that a reproducible cortical infarct could be produced in rats that had been administered the photosensitizing dye rose bengal prior to light irradiation. In this model, 560 nM light is delivered through the intact skull into the brain, where photochemical interactions give rise to microvascular injury and platelet aggregation.

Morphological and autoradiographic studies have shown that this particular protocol results in microvascular thrombosis, primarily within small pial and parenchymal arterioles. Subsequent BBB breakdown and vasogenic edema give rise to progressive microvascular depression and lesion formation. The ability to position a thrombotic infarct in predetermined cortical regions lends this model to examination of the functional consequences of infarction in selective neural systems and remote brain regions. A shortcoming of this cortical infarct model is that examples of small-vessel thrombosis are rare in clinical stroke. The ability to reproducibly infarct a predetermined cortical region has significant advantages in investigations that seek to determine the behavioural consequences of cortical infarction and the mechanisms underlying the recovery of function after stroke. For example, studies have documented the functional consequences of infarction of the forelimb, hindlimb, and posteromedial barrel subfield of the primary somatosensory cortex of rats.

In one study, unilateral and bilateral infarctions of the primary somatosensory cortex led to acute deficits in the ability of the rat to correctly run a "T" maze task. Interestingly, over a 1-month period, recovery of function in both unilateral and bilaterally infarcted rats was observed. In contrast, unilateral infarction of the hindlimb cortical region has been reported to lead to permanent disabilities in hindlimb placing tasks. These types of experimental strategies have led to new information concerning the mechanisms of recovery after focal brain infarction. For example, in chronically (2 months) barrel field infarcted rats, unilateral whisker stimulation combined with 2-deoxyglucose (2-DG) autoradiography has provided metabolic evidence for alternate patterns of somatosensory circuit activation. Interestingly, ipsilateral cortical areas adjacent to the infarct, as well as somatosensory cortical regions within the intact contralateral cortex, responded to whisker.

These experimental data indicate that intact regions may fake over function after cortical injury, a finding that has been supported by clinical investigations. Importantly, pharmacological agents that promote alternate circuit activation after focal brain injury may be useful in promoting functional recovery after stroke. Another advantage of the photochemically induced cortical infarction model is that the functional consequences of thrombotic infarction on remote, noninjured brain regions can be investigated critically.

The term "diaschisis" is used to describe the prolonged depression of neuronal activity in brain regions remote from the histopathological lesion. Investigations using the photochemical model have described remote depression of glucose utilization and ICBF after cortical thrombosis. Subsequent studies using the serotonin 5-HT_2 antagonist ketanserin have indicated that platelet-derived serotonin may participate in the remote hemodynamic depression after cortical thrombosis. Other studies have demonstrated repetitive episodes of cortical spreading depression (CSD), which move from proximal sites to areas more distal from the rim of the evolving infarct. Multiple episodes of CSD therefore may underlie some of the remote metabolic, consequences of thrombotic stroke.

The clarification of mechanisms underlying these remote responses is important in that therapeutic strategies that target these processes may reduce the initial behavioural deficit associated with stroke, as well as promote functional recovery.

Cerebral thrombosis

Thrombosis of the MCA has been shown to produce consistent infarcts within the MCA territory similar to those seen with mechanical occlusion. In one study, the topical application of nimodipine, a potent calcium channel, blocker and vasodilator, was reported to promote recanalization of the thrombosed MCA and lead to reperfusion. Direct comparisons between the histopathological consequences of transient MCA occlusion by mechanical, versus thrombotic methods have also been investigated.

Importantly, transient thrombosis led to more severe microvascular and neuronal consequences than transient mechanical occlusion of the MCA. These data indicate that thrombotic stroke may represent a more severe insult to the brain than cerebral ischemia alone. Another important use of animal models that incorporate large cerebral artery thrombosis is the study of thrombolytic therapy.

In previous studies, recombinant tissue, plasminogen activator (rt-PA) failed to lead to clot lysis of a photochemically induced MCA occlusion. In that study, the lack of fibrin within the platelet thrombus was suggested to explain this negative effect of rt-PA. However, subsequent studies from other laboratories using the photochemical method have presented positive data with rt-PA in clot lysis. Models where reperfusion can be successfully induced following periods of vascular thrombosis are important because thrombolytic agents including rt-PA are being used successfully in the clinic. In that setting, the potential for hemorrhagic transformation after successful thrombolysis is a serious concern. Thus, the ability to investigate, this issue in an animal model has important implications.

The thrombotic focal cerebral ischemia has been described in rats. In this model, a thrombus is induced at the origin of the MCA by the injection of thrombin via an intraluminal catheter placed in the internal carotid artery. Importantly, the thrombus generated by this technique is sensitive to treatment with rt-PA. Thus, this thrombotic model is relevant to the clinical problem of thrombolysis and may be useful in determining which factors predispose all ischemic zone to undergo hemorrhagic transformation after successful thrombolysis.

Cerebral embolization

Several experimental models have been developed to produce cerebral embolization in the brain. The injection of homologues blood clot fragments into the MCA of rats leads to a high incidence of ipsilateral infarcts. The injection of 35-µm carbon microspheres into the internal carotid artery of rats has also been shown to produce multifocal infarctions and the evolution of cerebral edema. Thromboembolic stroke has been induced in rats by using a photo chemical insult to the common carotid artery.

In this model, the CCA is irradiated following the administration of the photosensitive dye, rose bengal. Specific irradiation procedures lead to approximately 75% stenosis of the carotid artery. Because stenosis and not complete occlusion is produced, the carotid thrombus fragments after the irradiation period, resulting in a shower of platelet emboli into the distal cerebral microcirculation. This carotid thrombosis model has been shown to produce multiple embolic strokes, the majority of which are in the ipsilateral thrombosed hemisphere.

Interestingly, concomitant with platelet accumulation, blood-brain barrier breakdown at sites of platelet accumulation has been documented. Behavioural strategies for assessing sensorimotor and cognitive function have shown that this model results in behavioural deficits that are reminiscent of those seen in acute stroke patients. Transient platelet accumulation, leading to transient ischemic attacks, occurs in a large percentage of patients that ultimately develop an ischemic stroke. Abnormalities in cerebrovascular reactivity or hemodynamic reserve can be risk factors for stroke. Brief periods of systemic hypotension may be detrimental to the postinjured brain if autoregulatory mechanisms of cerebral blood flow are perturbed. Studies have determined whether the postthrombotic brain has an increased vulnerability to secondary insults that are normally noninjurious to the healthy brain.

In one study, the effect of moderate hypotension (60 mm Hg) induced 1 hr after CCA thrombosis was investigated. Hypotension induced after CCA thrombosis but not mechanical occlusion led to aggravated hemodynamic and histopathological damage The clarification of mechanisms underlying the increased sensitivity of the postthrombotic brain to secondary insults could lead to the development of novel therapeutic strategies that could limit brain damage in patients at risk for stroke.

Regulation of ischemic outcome

The major determinants of ischemic outcome are ischemic duration and severity. The question of whether complete or incomplete ischemia is worst in terms of neuronal survival was once a highly debated issue. Although some blood flow through the tissue would be expected to be better than no blood flow at all, excise lactate accumulation during incomplete ischemia has been reported to hinder brain recovery. This important question was answered by directly comparing the histopathological consequences of complete versus incomplete ischemia in a single rat model.

Under conditions that resulted in similar lactate levels, complete ischemia led to severe postischemic perfusion deficits, including no reflow and severe histopathological damage. In contrast, similar hemodynamic and histopathological consequences were not observed following incomplete ischemia.

Thus, ischemic severity plays an important role in the histopathological consequences of transient cerebral ischemia. Experimental data have demonstrated the importance of small differences in intra and postischemic brain temperature on histopathological an functional outcomes in various ischemia models. Indeed, the ability to dramatically modify the degree of ischemic damage by artificially varying intraischemia brain temperature represents a powerful tool by which to investigate ischemic pathobiology. Uncontrolled brain hypothermia commonly occurs during the ischemic period and that this consequence attenuates the degree of histopathological damage.

In one study where ischemic brain temperature was artificially controlled at specific levels during the ischemic period, mild reductions in brain temperature resulted in significant neuroprotection of selectively vulnerable brain regions, including the CA1 subsector of the hippocampus and dorsolateral striatum. Thus, it is extremely important in experimental studies of brain injury to monitor, and maintain brain temperature. Such, a procedure is important not only to promote lesion consistency but also to allow for clinically relevant pathological events to be active, in an experimental model.

Postischemic hypothermia is also neuroprotective following both global and focal ischemia. The specific method used to occlude a cerebral vessel may be important in terms of the clinical relevance of the injury model as well as the pathogenesis of the injury process. In one study, the neuronal consequences of recirculation following MCA occlusion induced either mechanically (surgical clip) or photochemically (platelet thrombus) was investigated.

This study clearly showed that the extent of brain damage after clip removal was significantly less than that seen after recanalization of the thrombosed MCA. The difference in histopathological outcome between the two models was attributed to microvascular abnormalities and reperfusion, injury. These results are important from the clinical standpoint in that the therapautic window for successful thrombolysis may be shorter than the duration predicted by results based on conventional models of transient cerebral ischemia. Thus, one must be careful when directly relating experimental data to the clinical condition of stroke. An experimental variable that may also significantly influence ischemic outcome is animal survival period. In previous studies, relatively short survival periods ranging from one to several days were used to monitor histopathological outcomes.

However, data indicate that histopathological processes may continue to evolve long after the primary insult. In one study that investigated a relatively short global ischemic period (10 min), significant neuronal damage within the dorsolateral striatum that wits not observed at 3 days, was present at 8 weeks of recirculation. A similar pattern of delayed neuronal damage has been reported in the striatum alter a brief period of MCA occlusion. In that study, cell death by apoptosis was suggested to underlie the delayed neuronal response to transient focal ischemia. The observation that the survival period is a significant factor in assessing the histopathological consequences of cerebral ischemia is also important from a therapeutic standpoint.

For example, it is possible that significant degrees of neuroprotection may be documented at short but not chronic survival periods. In this regard, a restricted period of immediate postischemic hypothermia (3 hr/ 30°C) has been reported to provide significant CA1 hippocampal protection at 3 days, but not at 2 months. Interestingly, most pharmacological studies in which drug treatments have been evaluated have used relatively short survival periods. From a clinical standpoint, it is important that chronic protection be demonstrated experimentally, prior to the initiation of clinical trials based on a particular therapy.

Pathophysiology

Several excellent review articles have been published that summarize the multiple pathological processes underlying ischemic cell death. Glutamate, the major endogenous excitatory neurotransmitter in the central nervous system has shows to mediate pathological central nervous system (CNS), has been processes in many injury models. In cell culture studies, Choi and Rothman demonstrated the neurotoxic effects of glutamate application and the importance of *N-methyl-D-aspartate (NMDA)* receptor activation in the injury process.

In animal studies, systemic injections of glutamate produce neuronal damage in brain regions lacking a BBB. Evidence for excitatory amino acids contributing to ischemic damage GAI hippocampus, can be protected by prior differentiation of glutamatergic pathways. With the use of intracerebral microdialysis, sampling of the brain's extracellular space has shown massive release of glutamate as well as other neurotransmitters during and following periods of global and focal ischemia.

In models of focal ischemia, the amount of glutamate release has been correlated with infarct size. Interestingly, intraischemic hypothermia reduces the levels of extracellular excitatory amino acids compared to normothermia. Most neurochemical studies using regional microdialysis have documented a significant rise in the extracellular levels of various neurotransmitters, including glutamate, during the immediate recirculation period. This acute rise followed by rapid normalization of neurotransmitter levels might indicate that the therapeutic window for protecting the brain against such agonist-receptor mediated events is relatively brief.

However, data from both experimental and clinical studies indicate that a secondary rise in the extracellular levels of glutamate can occur. Importantly, in clinical studies where levels of glutamate and other neurotransmitters are measured, elevated levels can be documented days after brain injury. Inflammatory processes occurring days after the primary ischemic period may potentially lead to secondary neuronal cell death by excitotoxic mechanisms.

For example, after transient global ischemia, postischemic hypothermia has been reported to provide only temporary protection of the CA1 hippocampus. One potential reason for the lack of chronic protection with postischemic hypothermia is that surviving neurons undergo a secondary insult that might originate from inflammatory processes including microglial activation.

On the basis of this hypothesis, the beneficial effects of combined therapy that included immediate postischemic hypothermia and delayed NMDA receptor blockage were investigated. Hypothermia in combination, with delayed MK-801 treatment led to chronic protection of the CA1 hippocampus. These studies suggest that the postischemic brain may undergo excitotoxic insults long after the primary injury and that multiple therapeutic windows may exist for postischemic neuroprotection. Free radicals are molecular species wherein the outer electron orbital contains only a single electron rather than the usual pair. These molecules are highly reactive and capable of inducing injury to lipid proteins and nucleic acids.

Investigations have implicated free radicals in ischemic brain injury. Lipid peroxidation, a method by which radicals are capable of damaging cell membranes, has been demonstrated in some but not all experimental studies. By using microdialysis techniques, local hydroxyl radical production has been documented following both global and focal ischemia studies using transgenic mice overexpressing human Cu, Zn-SOD, an endogenous enzymatic defence mechanism against oxidative damage have reported reduced in farcts compared to their noneransgenic litter mates. Calcium plays a critical role in cellular function, and various studies have implicated calcium ion in the, pathogenesis of ischemic brain injury.

During periods of cerebral ischemia, energy-dependent ionic pumps that are responsible for maintaining low intracellular calcium levels fail, resulting in the collapse of ionic gradients. In various brain injuries, intracellular free calcium has been shown to rise dramatically from both extracellular and intracellular stores. Thus, attempts, have been made to protect the ischemic brain from damage by blocking calcium entry during ischemia. In the brain, nitric oxide (NO) is involved in many biological processes, includirg circulation and neutral transmission. In pathological states, NO participates in neuronal excitotoxicity. Although much experimental data have been collected, the evolvement of NO in the pathophysiology of brain injury remains controversial.

Inhibitors of nitric oxide synthase (NOS) have been used to modulate NO formation in models of global and focal ischemia. Thus NOS inhibition has been reported to be beneficial in some experimental settings, whereas in others deleterious consequences have been reported. The infusion of L-arginine to augment NO production or the administration of NO donors has been reported to be neuroprotective in some models. The use of neuronal and endothelial NOS knockout mice has identified the importance of specific NOS isoforms in ischemic injury.

Animals deficient in neuronal NOS appear to be more resistant to ischemic tissue damage than the wild type. The importance of NO in the pathophysiology of thrombotic stroke hag also been investigated. In a model of MCA thrombosis, Prado and colleagues failed to demonstrate a beneficial effect of L-arginine administration on the early hemodynamic pertubations or histopathological consequences of MCA thrombosis. These data contrast with positive findings from studies that used more conventional ischemia models. The majority of ischemic strokes result from thromboembolic processes. It is therefore important to determine the role of NO in the pathophysiology of this condition. Photochemically induced common carotid artery thrombosis (CCAT) with subsequent platelet embolization to the brain has been used to investigate this question.

Stagliano and colleagues reported that limiting endothelial NOS activity aggravated the histopathological and behavioural consequences of CCAT, whereas the inhibition of neuronal NOS activity and the i fusion of an NO donor n improved outcome. Interventions to restore cerebral blood flow to an ischemic region would benefit the brain, experimental and clinical evidence indicates that organ reperfusion may exacerbate injury. Thus, as a paradoxical consequence of postischemic reperfusion, injury may occur. Experimental data first obtained in the heart, lung and kidneys indicate that reperfusion of ischemic tissue resulted in irreversible injury that could not be attributed to the ischemic insult alone. In stroke patients, fatal edema formation during successful thrombolysis by septokinase or rt-PA has been reported. Experimental studies have described a wide spectrum 4 vascular responses to periods of cerebral ischemia that would be expected to influence outcome.

Alterations in blood-brain barrier permeability have been reported following global and focal ischemia. The breakdown of the BBB may aggravate ischemic out come by multiple mechanisms, including increased edema formation and extravasation of blood-borne factors that could promote vascular and neuronal damage. Increased brain water would increase intracranial pressure and lead to vascular compression and decreased tissue perfusion. Various studies have documented the temporal profile of BBB alterations and attempted to clarify pathomechanisms underlying these vascular perturbations. Although the BBB remains intact for several hours during permanent ischemia, acute alterations in barrier function occur with recirculation.

Interest has been directed toward the question of whether white blood cells (WBCs), particularly polymorphonuclear leukocytes (PMNL), could act as mediators of secondary brain damage in models of cerebral ischemia. The temporal profile of PMNL accumulation therefore has been documented in various ischemia models. Hallenbeck and colleagues first documented significant PMNL accumulation at 4 hr after embolic ischemia. Subsequent studies have reported that PMNL accumulation appears to be maximal at 24-48 hr after injury. Studies where the induction of neuropenia was pharmacologically induced reported improved outcomes.

WBCs may plug obstruct microvessels, release vasoconstrictive mediators, or migrate into the.,brain tissue where they may damage neurons by the release of cytotoxic enzymes or free radicals. These studies have stimulated research into the mechanisms underlying, WBC-mediated tissue injury. Adhesion of PMNL to the endothelium requires the activation of various adhesion receptors on WBCs and the vascular endothelium. For example, the β_2- integrin CD11/CDI8 has been identified on WBCs and together with ICAM-1 on enclothelial cells, is involved with WBC accumulation and endothelium. Methods to block specific adhesion receptors have been developed to inhibit WBC accumulation and potentially lead to neuroprotection.

Majority of ischemic neurons are believed to die by necrotic mechanisms, data indicate that apoptotic cell death may also contribute to the ischemic lesion. Evidence supporting a role of programmed cell death in focal ischemia was first provided by Linnik and colleagues, who reported that new protein synthesis appeared to contribute to ischemic injury in which DNA degradation was apparent. In that study, treatment with the protein synthesis inhibitor cycloheximide reduced infarct size after permanent MCA occlusion. Although which putative proteins are important to apoptotic cell death are unknown, several, including immediate early gene products, heat shock proteins, ornithine decarboxylase, and p53, increase folio wing an ischemic insult.

Thus, many studies have been published that provide, morphological and biochemical data for apoptotic cell death occurring in various ischemia models. A biochemical index for apoptosis is the oligometric laddering of DNA, caused by internucleosomal cleavage. The presence of laddered DNA fragmentation using gel electrophoresis has been reported in models of global and focal ischemia. In contrast, smeared DNA fragments, consistent with the frank necrosis, result from various acute ischemic injuries.

The TUNEL method has been used extensively as a morphological method of detecting DNA fragmentation and apoptotic bodies. However, limitations regarding the use of this stain to determine apoptosis have been reported. Because necrotic neurons may also diffusely stain with the TUNEL method, alternative approaches need to be utilized. Studies have relied upon several different approaches, including ultrastructural analysis, to document apoptotic cells. Morphological data are needed to determine which specific cell types (i.e., neurons, glia, leukocytes, or macrophages) undergo apoptotic cell change after an ischemic insult and how these cellular responses influence ischemic outcome. To determine which therapeutic interventions may be effective in the postischemic condition, future studies will also need to document the percentage of neurons undergoing necrosis versus apoptotic cell death.

Neuroprotection

The models of transient forebrain ischemia with subsequent recirculation give initial reports indicated neuroprotection with MK-801. Subsequent studies, however, demonstrated a lack of significant protection of the CA1 hippocampus. An important observation indicated that prolonged hypothermia induced by MK-801 treatment may be responsible for the neuroprotection seen in some experimental models.

Other data indicate that the postischemic blockage of AMPA, but hot NMDA, protects the CA1 hippocampus from a global ischemic insult. In contrast to global ischemia, focal ischemia appears to be responsive to NMDA antagonism. Focal ischemia differs from global ischemia in that it has a therapeutically susceptible zone of the penumbra where the potential for collateralization exists. MK-801 administration has been shown to abolish repetitive ischemic depolarizations in the peri-infarct region which have been hypothesized to contribute to penumbral vulnerability. In this setting, repetitive episodes of CSD might lead to sustained metabolic-blood flow uncoupling, a condition that may induce penumbral damage. Other noncompetitive NMDA antagonists, including dextromethorphan, CNS1102, and SL82.0715 (eliprodil), have neuroprotective effects in models of focal ischemia. In addition to noncompetitive NMDA antagonists, competitive NMDA antagonists have been reported to be neuroprotective in various ischemia models.

In global ischemia, the competitive NMDA antagonists, CGS19755 and CPP, provide neuroprotection when given in the preischemic condition. In focal ischemia CGS19755, given before and after vascular occlusion, also decreased infarct volume. CGS19755 is currently being tested in a stroke clinical trial. In addition to blocking the NMDA receptor, other therapies are directed toward inhibiting glutamate release.

The compound BW1003C87 has been shown to inhibit the veratrine-induced glutamate release. The compound also attenuates extracellular elevations in glutamate after ischemia and protects the CA1 hippocampus following global ischemia.Calcium channel antagonists have been shown to be neuroprotective in both global and focal cerebral ischemia models.

However, most studies evaluated preischemic treatment, and a lack of significant protection when given after MCA occlusion has commonly been reported. For example, continuous treatment for 24 hr after MCA occlusion in baboons has been shown to improve cerebral blood flow (CBF) but not significantly decrease infarct volume. The calcium channel blocker flunarizine has been reported to improve behavioural function after cortical infarct formation, but failed to affect infarct volume. (S)-Emopamil, averapamil-like calcium channel blocker and serotonin 5-HT_2 antagonist, has been shown to decrease infarct volume when initiated I hr after permanent MCA occlusion in rats. The importance of oxygen radicals in the pathobiology and treatment of cerebral ischemia has been reviewed. The enzyme superoxide dismutase (SOD) conjugated to polyethylene glycol (PEG), which increases the circulation half-life of SOD by many hours, reduces infarct volume in, permanent MCA occlusion by approximately 25%. Transgenic mice overexpressing human Cu,Zn-SOD have been developed to study cerebral ischemia.

Under conditions of permanent MCA occlusion, transgenic animals have redueed infarct volumes compared to those of their litter mates. In subsequent studies, however, following transient MCA occlusion transgenic mice, demonstrated more profound reductions in infarct volumes. Thus, antioxidant therapy appears to be most beneficial in experimental conditions that,include reperfusion. A Phase II trial with PEG-SOD in patients with severe head injuries showed that a high dose seemed to improve outcome, whereas a Phase III trial showed no significant effect. Compounds that inhibit lipid peroxidation provide neuroprotection in various experimental ischemia, models. A series of 21 aminosteroid compounds by Upjohn, including U74006F, protec'ed hippocampal CA1 neurons following global ischemia in gerbils.

In focal ischemia, dose-dependent reductions in infarct volume have been reported in, rats with early administration of tirilizad. A randomized trial, of high-dose tirilizad has been terminated in Europe and North America because of interim results demonstrating a lack of benefit in acute stroke. Nitric oxide synthesized from L-arginine by a family of NO synthase (NOS) enzymes has been, studied extensively in models of cerebral ischemia. There are at least three isoforms of

NOS. The endothelial and neuronal forms are constitutive and are whereas the macrophages contain all inducible form.

Experimental studies have shown that the systemic administration of NOS tends to reduce resting CBF and that NOS inhibitors attenuate hypercapnea-induced CBF increases. Studies of NOS inhibition of cerebral ischemia have suggested both beneficial and detrimental effects of this compound. In neuronal cultures, NO contributes to the NMDA-mediated toxicity of neurons, possibly by radical-mediated mechanisms. The use of nonselective pharma-cological NOS inhibitors in focal cerebral ischemia has produced contradictory results. These controversial results may reflect different degrees of inhibition of endothelial versus neuronal NOS.

Administration of the NO precursor, L-arginine, has been reported to increase penumbral CBF and reduce infarct volume. Work on mutant mice deficient in the neuronal form of NOS has demonstrated decreased infarct volume after MCA occlusion. In contrast, infarct volume in mutant mice was larger when endothelial NOS was inhibited by nitro-L-arginine. Several growth factors have been reported to be neuroprotective in models of neuronal injury. The neurotrophin family of growth factors include brain-derived neurotrophic factor (BDNF), nerve growth factor (NGF), and neurotrophins 3 and 4.

NGF has been reported to protect neurons from transient global ischemia, whereas BDNF administered before global ischemia promotes the survival of CA1 hippocampal neurons. In one study, BDNF given to rats 2 days before and 7 days after global forebrain ischemia also decreased the frequency of ischemic CA1 neurons. Basic fibroblast growth factor (bFGF), an 18-kDa polypeptide, exerts trophic effects on neurons, glia, and endothelium and is capable of protecting neurons in culture from excitotoxicity and other neurotoxic influences. Intraventricular and intravenous administration of bFGF following MCA occlusion has been shown to reduce infarct volumes. Finally, bFGF administration, starting 24 hr after MCA occlusion, has been reported to accelerate behavioural recovery after stroke without affecting chronic infarct size. Thus, in addition to being neuroprotective, bFGF may also improve behavioural recovery following complete stroke.

Various experimental studies have implicated roles for several inflammatory molecules in the pathophysiology of stroke. These include tumor necrosis factors-α, interleukin-1β, and interleukin-6 as well, as various chemokines. Cytokines are believed to be cytotoxic to brain cells and may also serve as chemoattractants for blood-borne leukocytes. Cytokine-suppressive drugs therefore have been suggested as a possible stroke therapy. The intracerebroventricular administration of IL-β significantly exacerbates neuronal loss after transient global ischemia. Several reports have been published regarding the neuroprotective effects of human re combinant interleukin-1 receptor antagonist (rhIL-1ra) in ischemia models.

In a model of MCA occlusion, Relton *et al.* reported that rhIL-1ra treatment markedly reduced infarct volume in rats. These promising data indicate that pharmacological strategies that inhibit the synthesis or actions of cytokines, including IL-1, may offer a benefit in the treatment of stroke. In the heart and brain, ischemic preconditioning (IPC) is defined as the protection seen when mild, sublethal ischemic episodes decrease dam age from subsequent lethal ischemic insults. In the heart, IPC protects from subsequent sustained coronary occlusion both in the region in which IPC was elicited and in remote myocardium. In the brain, histopathological damage has also been shown to be less when a sublethal insult precedes a lethal insult by 1 or 2 days. IPC protection depends on the intensity of the ischemic insult and the latency between the two insults. In cardiac muscle, for example, the maximum latency between the insults that is necessary for IPC protection is 1 hr. In the brain, however, IPC paradigms are less well-defined. IPC in the brain has been reported when many hours of reperfusion separate the conditioning and test insults. Data indicate that rapid preconditioning can also be observed following transient cerebral ischemia.

Perez-Pinzon and colleagues reported that a preconditioning ischemic, insult of 2 min associated with anoxic depolarization significantly decreased the frequency of CA1 hippocampal ischemic neurons when the second lethal insult was induced 30 min later. The mechanisms by which pretreatment with a sublethal injury, whether global or focal, induces resistance to a subsequent longer period of ischemia are not known. The possibility that cortical spreading depression (CSD) may have a neuroprotective action during subsequent focal ischemia has been examined. Thus, artificially induced evoked CSD has been shown to induce ischemic tolerance after global and focal ischemia.

On the basis of experimental data showing neuroprotection in various animal models of cerebral ischemia, clinical trials have been initiated in stroke patients. Although more than 35 randomized controlled trials of various neuroprotective strategies have been reported since the mid-1980s, a successful treatment has finally been demonstrated with the thrombolytic agent rt-PA. Stroke represents an heterogeneous population of patients and therefore presents a potential problem in targeting which specific patients might benefit from pharmacological intervention.

In addition, because the therapeutic window for treatment is limited, successful treatment frequently depends on whether the patient arrives at the emergency room soon after the first signs of neurological dysfunction. Although measures that induce clot lysis and restore cerebral recirculation may limit the degree of ischemic damage, the potential for intracranial bleeding is of major concern. Several clinical studies have described the results of systemic or intra-arterial administration of various thrombolytic drugs. In the NINDS rt-PA stroke study that included 624 patients., intravenous -rt-PA improved outcome after stroke when given early and within 3 hr of onset. Thus, if treatment can be initiated by 3 hr post-onset, thrombolytic therapy with rt-PA produces a beneficial outcome irrespective of age, gender, ethnicity, or presumed cause of stroke.

Nevertheless, thrombolytic therapy should not be given unless facilities to handle bleeding are available. In contrast to the findings with rt-PA, intravenous administration of the thrombolytic agent streptokinase outside the clinical investigation was not indicated. Subsequent investigations have attempted to find additional information that might be used to improve patient selection for rt-PA therapy after stroke. In one study, various stroke risk factors including age, history of, diabetes, hypertension, or early CT findings were examined. However, no baseline patient data could be identified that predicted successful outcome. However, other data indicate that early CT signs of ischemia and severe neurologic deficits at baseline were associated with a greater risk of symptomatic intracerebral haemorrhage after rt-PA therapy.

On the basis of promising experimental data-several clinical studies have been initiated to test the efficacy of *N*-methyl-D-aspartate (NMDA) receptor blockers in acute stroke. Some studies have been stopped prior to completion due to a low projected likelihood of success and an increased incidence of mortality in treatment groups. A phase III efficacy and safety study is underway on the novel noncompetitive NMDA receptor blocker CNS1102. Patients receive drug treatment within 6 hr of onset of their ischemic stroke with the primary efficacy endpoint being the modified Rankin Scale at day 90.

Lubeluzole is a novel neuroprotectant that in experimental studies improves functional outcome and minimizes tissue damage after focal ischemia. This agent interferes wit the toxic effects of nitric oxide and reduces extracellular levels of glutamate following infarction. In a double-blind Phase II trial of 232 patients, lubeluzole infusion within 6 hr of stroke onset was shown to be safe and to significantly reduce mortality in the North American trial. However, data from the European trial failed to show a significant effect of lubeluzole treatment on mortality rate.